Critical Condition

Critical Condition

Human Health and the Environment

A Report by Physicians for Social Responsibility

edited by
Eric Chivian, M.D.
Michael McCally, M.D., Ph.D.
Howard Hu, M.D., M.P.H., Sc.D.
and Andrew Haines, M.D.

The MIT Press
Cambridge, Massachusetts
London, England

This book was set in Bembo by .eps Electronic Publishing Services and was printed and bound in the United States of America.

Library of Congress Cataloging-in-Publication Data

Critical condition : human health and the environment / edited by Eric Chivian . . . [et al.].
     p.   cm.
  Includes bibliographical references and index.
  ISBN 0-262-03212-0 (hc).—ISBN 0-262-53118-6 (pbk.)
  1. Environmental health. 2. Pollution—Health aspects. 3. Environmental degradation—Health aspects. I. Chivian, Eric. II. Title: Human health and the environment.
  [DNLM: 1. Environment. 2. Public Health. 3. Environmental Health. WA 30 C934 1993]
  RA565.C75   1993
  616.9′8—dc20
  DNLM/DLC   93-17760
  for Library of Congress   CIP

He [modern man] commonly thinks of himself as having been here since the beginning—older than the crab—and he also likes to think he's destined to stay to the bitter end. Actually, he's a latecomer, and there are moments when he shows every sign of being an early leaver, a patron who bows out after a few gaudy and memorable scenes.

—E. B. White, *Second Tree from the Corner* (1936)

If you make an enemy of the Earth, you make an enemy of your own body.

—Mayan shaman

Health professionals should take the lead in moves to improve the environment and to inform governments and the public. . . .

—Report of the World Health Organization Commission on Health and the Environment (1992)

This work was supported in part by the Frank Weeden Foundation.

Contents

Preface

Since 1961, Physicians for Social Responsibility (PSR) has been leading an effort to educate the public about the medical consequences of nuclear war, driven by the conviction that there was little under standing, even at the highest levels of government, of what would happen in a nuclear war, and that this lack of understanding made a nuclear war more likely to occur. This state of affairs, it was felt, constituted an enormous danger to human health and survival.

The educational programs of PSR physicians were based on the belief that if people really understood the effects of nuclear weapons, in the concrete, personal terms of human health, they would no longer think it was possible to fight and survive a nuclear war, and would demand an end to the nuclear arms race. It was for these efforts that PSR shared the 1985 Nobel Peace Prize with its international umbrella organization, International Physicians for the Prevention of Nuclear War.

The world now faces a similar threat to human health and survival from changes to the global environment—stratospheric ozone depletion, habitat destruction, species extinction, global warming, and the poisoning of air, water, and soil by toxic and radioactive substances. And there is a similar lack of understanding about the consequences of these environmental dangers for human beings.

Tackling environmental degradation, however, is a far more difficult task. The problem is more complex by orders of magnitude than that of nuclear war, and its solution will demand much greater changes in the way people lead their lives, in developed and developing countries alike.

We believe that people will be motivated to make these changes, but only when they have grasped the full dimensions of the environmental crisis—namely that their health and lives, and those of their children, are at stake. That is why we have assembled this volume.

The idea for the book grew out of discussions between two of the editors (E.C. and M.M.) and Dr. Noel Brown, Director for North America of the United Nations Environment Programme, in mid 1991 and early 1992. A need was identified at these meetings for a comprehensive and widely available report of what was known about the human health effects of environmental degradation—a subject that, despite its obvious importance, has been given little attention in environmental discussions.

Such a report, it was decided, had to gather in one volume the best available medical information about the environment. It had to be comprehensive and technical enough to satisfy the needs of a well-trained physician or scientist, yet clear and understandable to the general reader. It had to cover not only those topics traditionally covered in standard courses and texts of public health—air and water pollution, food contamination, occupational health—but also environmental threats that have often been neglected or that are new areas of inquiry, such as nuclear power, war and military preparation, global climate change, stratospheric ozone depletion, and biodiversity loss. Above all, it had to establish that population growth stands at the center of global environmental problems, and that all attempts to solve those problems will fail until the exponential growth in the world's population is halted. All these things we have tried to do in the present work.

The role of physicians is given much attention in this volume. We hope they will become central figures in environmental policy discussions, as they have been for the issue of nuclear weapons. We also hope that the materials which follow will help environmentalists become more knowledgeable about environmentally induced illness, so that their advocacy for environmental protection will be even more effective. No arguments for the environment are more persuasive than those made from the perspective of human health. Finally, we hope that this volume will make the case that it is essential for the medical and environmental communities to work together.

Many people, directly and indirectly, helped with this book. In particular, we are indebted to those scientists whose work forms the intellectual underpinnings for our own—Irving Selikoff, Alice Hamilton, Sir Richard Doll, Arthur Westing, Paul Ehrlich, E. O. Wilson, Richard Schultes, Rachel Carson, Roger Revelle, Paul Crutzen, Stephen Schneider, Alice Stewart, and Sherwood Rowland, to name a few.

We also owe a debt to indigenous peoples around the world, from whom we have much to learn, not only about native flora and fauna that contain valuable medicines, but also about how to live in harmony with one's environment. It is fitting that this volume is being published in 1993, the United Nations' Year of Indigenous Peoples.

Many others have assisted with this volume—the Frank Weeden Foundation, which believed in the value of the project from the time it was first proposed and provided startup funds; Frank Urbanowski and Madeline Sunley of The MIT Press, who strongly supported the idea for the book and worked to see it happen; and Cristina Sanmartín, also at The MIT Press, who prepared the manuscript. To them and the many others who helped along the way, we are deeply grateful.

Finally, we want to call special attention to the roles played by Dr. Noel Brown and Vice-President Albert Gore. Dr. Brown's challenge to us to produce a report on human health and the environment provided the initial spark for this project, and his consistent support over two years' time moved it forward. Vice-President Gore expressed continued interest in our work from the beginning, and has provided for us a model of moral courage, scientific curiosity, and intellect rarely found in public figures. We hope this volume will be a resource both at the United Nations and with the new American administration for understanding the medical consequences of environmental degradation, and will be seen as a physicians' warning that there is still time to preserve the environment and protect human health.

Eric Chivian, M.D.
Michael McCally, M.D., Ph.D.
Howard Hu, M.D., M.P.H., Sc.D.
Andrew Haines, M.D.

Introduction: Human Health, Risk, and the Environment

Anthony D. Cortese, Sc.D.

Three important themes underlie this book:

- The physical environment, our habitat, is the most important determinant of human health.
- Protection of the environment and preservation of ecosystems are, in public health terms, the most fundamental steps in preventing human illness.
- Physicians should be the health officials most knowledgeable about the environmental factors that cause disease, and should be prominent spokespersons in communicating with the public about environmental hazards.

Environment Determines Health

We have known for centuries that a healthy environment is essential for human existence and health, and that contamination of the environment with heavy metals, microorganisms, physical agents, and certain organic compounds can cause serious illness and death. Improved water, food, and milk sanitation; reduction in physical crowding; better nutrition; and central heating with cleaner fuels were the intervention strategies most responsible for the marked improvements in public health that were achieved during the 20th century.

Environmental health problems, which had been local in their effects and short in their duration, have changed dramatically in the last 20 to 30 years. Many of today's environmental problems, such as acid rain, high urban ozone pollution, and pollution of the Mediterranean Sea, are both regional and international in scope. In the United States alone, 100 million people breathe levels of ozone dangerous to health at various times of the year. Acid rain is causing damage to forest and aquatic ecosystems on every continent. Depletion of the stratospheric ozone layer, which protects us from harmful solar ultraviolet rays, and changes in global climate due to the emission of "greenhouse gases" into the atmosphere have the potential to affect virtually every human being on the planet. These long-term environmental problems are not amenable to quick technological fixes. The effects often occur many years in the future and are highly uncertain.

Environmental Problems Are Global and Long Term

For the first time in human history the cumulative effects of rapid increases in population and even more rapid industrial growth, with the attendant pollution and resource depletion, are causing changes on a global scale.[1,2] We are altering the basic physiology of the planet—the chemical composition of the atmosphere and the cycling of carbon, nitrogen, and phosphorus (the basic elements of life) through the biosphere. Tropical forest destruction, soil erosion, the withdrawal of water from the hydrologic cycle, and toxic pollution are causing ecological changes which threaten the quality of life and indeed the survival of humanity. The threatened destruction of this natural support system will be one of the most important international issues to be faced by society and by the public health community in the 21st century.

The rate of industrialization, which is far outpacing the growth in population, is an even more powerful determinant of environmental transformation. In the past 100 years, the world's industrial production increased 100-fold.[2] From 1950 to 1985 manufacturing increased by a factor of 7, the use of electricity by a factor of 8, the number of automobiles by a factor of 7 (from 50 million to 350 million), and the production of synthetic chemicals by several orders of magnitude.[3] The impact on the global environment has been dramatic. In 150 years human activity has increased the atmospheric concentration of carbon dioxide by 24%, doubled the concentration of methane (both potent greenhouse gases), and introduced long-lived ozone-destroying chlorofluorocarbons into the stratosphere. Man-made emissions of sulfur and nitrogen, which lead to acid rain, now equal or exceed the natural flux of these elements. And man-made emissions of lead, cadmium, and zinc exceed the natural flux by a factor of 18.[2]

Human Belief Systems Are Part of the Problem

Fundamental human beliefs have produced this environmental transformation. In the anthropocentric view of the world, man is the most important of all the species and should have dominion over nature. Most humans believe that any harm we do we can undo, with ingenuity and technology, and that our individual and collective impact on the environment will result in only imperceptible changes.

We treat our natural resources as free and inexhaustible. There is little understanding of the concept of the "carrying capacity" of the environment. The frontier mentality—especially in the United States, with its vast natural resources—has led to the rapid consumption of resources. Tropical rain forests, which are thought to contain over 50% of the world's biological species, are being cut down at a rate of approximately one acre per second. Every year the world loses an area of rain forest about the size of Austria. Nearly 1.5 billion people are cutting down forests for fuelwood and agriculture at a nonrenewable rate.[3]

We believe that the environment has a nearly infinite capacity to assimilate any toxic discharge or waste. This belief has led to pollution-control strategies such as tall stacks to disperse local sulfur dioxide emissions to distant areas, and to the dumping of radioactive and other toxic wastes in the ocean. Finally, we have concentrated on simplistic, quick, technological fixes for environmental problems. These are often impractical, and they do not deal with the root causes of the problems. When large concentrations of smog appeared in Los Angeles, for example, one of the first proposals was to build large fans in the San Bernardino Mountains to blow the polluted air out to sea.

Solutions

Given the close connection between environment and health and the fundamental belief systems that prevent us from recognizing the dimensions of the environmental crisis, how can we promote health?

A first step is to stop further environmental degradation.[2] This will require major shifts from policies that control pollution and repair damage to policies that seek to prevent the generation of pollution in the first place.

A second step will be to change the relationship between developed and developing countries. Industrialized nations will need new strategies for transferring technologies, training, and education and for providing financial assistance to non-industrialized nations. These strategies must deal with international debt, and must promote economic development that minimizes the destruction of resources and the generation of pollutants while also improving the quality of life. The implementation of such "sustainable development" strategies will require a profound understanding by the developed world of the interdependence of all nations when it comes to the global environment.[4]

A third initiative is to develop long-term educational strategies that will change the mindsets of individuals and institutions with respect to protecting the environment and promoting health. The leaders of the 21st century must understand that virtually every human activity affects the health and welfare of the planet. Air, water, land, and plant and animal life are the bases of human existence. Physicians, engineers, business people, scientists, economists, politicians, and all other professional disciplines must understand environmental issues in relation to health and the quality of life, and must bear individual and collective responsibility for stewardship of the world's resources. Shifting our way of thinking to the interdisciplinary, the intergenerational, and the international will demand incorporation of these concepts into the curricula of all disciplines at the undergraduate and graduate levels.

Physicians' Skills

In answering questions about toxic exposure or other environmental health risks, a physician needs to call on a number of skills, including the ability to evaluate and assess the nature and the extent of the patient's exposure to an environmental hazard, the ability to assess the degree to which the hazard presents a health threat, and the ability to communicate the extent of the risk to the patient or to the community so as to minimize or prevent further exposure.[5]

One of the challenges in assessing an environmental exposure is getting some idea of the actual dose the patient received. The exposure-disease model involves four elements: the exposure to the substance, the dose the patient actually absorbed, the biological effect of the absorbed dose, and the clinical disease that results.[7]

Medical students receive little training in these skills. One study that examined the teaching of occupational and environmental medicine concluded that current medical students receive less than 4 hours of education in these two areas during their 4 years of medical school.[6] A decision must be made as to where in the medical school curriculum environmental health and risk assessment should be taught. The most likely candidate is the course on epidemiology (the study of the causes of diseases in populations), but in general this subject is given little attention. The same situation applies to the education of physicians after medical school—environmental and occupational health are rarely covered.

The term *exposure* refers to the intensity and duration of contact with a substance or with a physical agent such as ultraviolet, x-ray, or microwave radiation in air, water, food, or soil. Inhalation, skin absorption, and ingestion are the possible routes of exposure. How much a person actually absorbs from an exposure constitutes the *dose,* the most important determinant in clinical disease. What dose someone was exposed to depends on the amount, route, and duration of exposure and on the concentration and chemistry of the substance or the intensity and type of the physical agent. (The absorbed dose is generally greater if the exposure level is high and, in the case of chemical substances, if the substance involved is non–ionized and highly soluble in fat.) The biological effect depends on a substance's chemistry or a physical agent's type, on the dose, and on the patient's susceptibility to disease. A brief high exposure, for example, typically produces acute disease only.

Risk and Risk Assessment*

A *risk* is a potential threat to health and life. How we assess, perceive, communicate, and manage risks is termed *risk analysis*. A number of contentious questions arise in each of these areas. This is a substantive, changing, and controversial field. Although great progress has been made in the last 15 years, the science of risk analysis is the least well understood aspect of the health effects of pollutants and environmental change.

Patients worried about exposure to environmental or occupational agents often want to know what their risk of developing a disease will be. Satisfactory answers are hard to come by. Numbers have been derived from risk assessments; these numbers, however, are probability figures that apply to populations, not to individuals. And, though derived from factual data, the final numbers involve a number of assumptions and offer only an estimate of how much an exposure to a chemical or physical agent is likely to affect health in a population.[8]

It is important to remember that a final risk figure is a probability, not something that is guaranteed to happen. Consider, for example, the probability of developing lung cancer from smoking cigarettes—a case in which there is a clear-cut cause-and-effect relationship between exposure and disease. One cannot say unequivocally that any one smoker will develop lung cancer. All that can be said with certainty is that the smoker's risk, or probability, of developing lung cancer is increased by some amount.

Where do risk numbers come form, and how accurate are they? They are derived by a process called *risk assessment,* which has four components:

---

*The material on risk and risk assessment in this section and in the next three sections has been adapted, with permission, from "Assessing and explaining risk," in *Environmental Issues in Primary Care* (© 1991 Minnesota Department of Health).

hazard identification, which involves identifying health hazards in the environment and characterizing their physical and chemical properties, their environmental and biological fates, and their abilities to cause disease; *exposure assessment,* which involves estimating human exposure from air, water, food, or skin contact with the hazardous agent; *dose-response assessment,* which involves characterizing the potential of a substance to cause disease as a function of the exposure or dose; and *risk characterization,* which integrates information on the agent's toxicity, likely routes of exposure, and dose-response assessment into an estimate of the agent's probable effect on the health of a population.

The data used in deriving risk assessments come from clinical studies; from epidemiologic studies of disease in human populations; from animal studies, in which populations of laboratory animals are exposed to large doses of the agent; from *in vitro* tests of mutagenicity; from structure-activity relations, which identify potential toxic agents from their similarity to others that are known; from physiological, biochemical, and toxicological findings; and from comparative metabolic and pharmacokinetic studies. An assessment may weigh the risks of cancer, birth defects, neurotoxicity, immunological changes, or other health effects. To date, most risk assessments have focused on cancer risks, but future risk assessments will place more emphasis on immunological and neurological risks and on teratological risks (i.e., risks of fetal damage).[8,9]

Animal Studies

Clinical and epidemiological studies offer the most direct means of assessing human health risks associated with hazardous exposures. It is impossible in most cases to carry out on humans the kinds of experimental studies needed to make reliable risk predictions. As a result, many epidemiological studies are done on populations of humans who were exposed to a chemical or a physical agent accidentally or in the course of their work but whose exposure may be difficult to assess. In the case of a chronic disease, such as cancer, the long time between an exposure and the appearance of clinical disease makes it difficult to determine the agent's effects on health. In other cases, although a chemical's toxic effects may be apparent, lack of adequate exposure data may make it difficult to use available epidemiologic studies in quantitative risk estimates. Hence, data from laboratory studies usually form the backbone of risk assessment.[9]

Toxicologists readily admit that laboratory animals differ from humans in a variety of ways: humans are larger; live longer; frequently suffer from two or more diseases at a time; may receive very different exposures than lab animals do; may metabolize, store, or excrete the agent differently; and, as a species, are more genetically heterogeneous. Nevertheless, most toxicologists believe, on the basis of experimental evidence, that laboratory animals and humans have more physiological, biochemical, and metabolic similarities than differences.

Extrapolating Results to Humans

One objection to the use of animal data is that lab experiments typically use high doses so that the test will be sensitive and unlikely to give a false negative result; the risk estimate is then extrapolated from results seen at high experimental doses to the much lower levels—often 1,000 to 10,000 times lower—that are typical of environmental exposures. The low-dose estimate is then scaled up to humans, either by a "safety factor" approach or by mathematical modeling.

Many risk estimators prefer to use mathematical models to characterize the underlying dose-response curve. In carcinogenesis, risk estimators typically use one of three classes of models. *Tolerance distribution* models assume that each individual in a population has a threshold below which he or she will not respond, and that thresholds vary among individuals. *Mechanistic* models, based on the assumption that cancer originates within a single cell, attempt to characterize the process of transformation mathematically. *Time-to-tumor* models are occasionally used in low-dose extrapolation when data on the time between initial exposure and detection of a tumor in an individual animal are available. The model used can have a significant effect on low-dose risk estimates.[9]

Identifying developmental or reproductive toxic agents in humans is possible but is even more complex than identifying carcinogens. The risk assessment must characterize the timing of the exposure as well as the type and severity of the effects. The reproductive risk assessment is complicated by the fact that a high enough dose of virtually any substance, administered to the mother at the correct time, will have some kind of adverse developmental effect on the fetus.

Refining Risk Assessment

As a science, risk assessment is going through a period of rapid development. Exposure assessment, thought to be one of the most challenging steps in the process, is being refined to consider environmental degradation and accumulation, ingestion of contaminated soil and foliage by cattle, deposition rates of particulates, dermal uptake, biomagnification, pharmacokinetics in humans, exposure schedules, and the dietary changes that take place over a lifetime

New models that try to characterize the exposures of subgroups of a population may replace models that emphasize worst-case scenarios. Risk assessment is also beginning to emphasize, as was mentioned above, reproductive and developmental toxicity, neurotoxicity, and immunotoxicity as well as carcinogenicity. And finally, new pharmacokinetic models for interpreting rodent and human data may help explain differences in biological response between humans and other animals.[9]

Closely linked to risk assessment and risk perception is *risk communication:* the way(s) in which the existence and the relative magnitude of a hazard are communicated. This is too often left to regulators and politicians who know little about health and disease or to health professionals who know little about risk assessment.[10,11,14]

The Role of the Physician

An important target of environmental education ought be physicians. Most physician training is designed around the medical model of "finding and fixing" the health problem. We must change the orientation of our thinking toward creating health, rather than curing disease. This will require a shift from a bioengineering to an ecological model in medicine.[12] All physicians (especially primary-care physicians) need to understand the relationship of environment to health. They should be able to detect and diagnose environmentally related disease, know how to obtain information about environmental hazards, advise patients on intervention strategies to reduce exposure to environmental hazards, and be able to refer patients to specialists in environmental and occupational medicine.

Most physicians lack an adequate background in statistical reasoning and in environmental and occupational epidemiology, and many risk assessors have not been adequately grounded in human disease biology or in toxicology. There are very few physicians in the U.S. Environmental Protection Agency, and it has been argued that "what is needed is a concerted effort on the part of EPA to hire individuals educated in public health and to provide public health training to present employees."[13]

It is time for physicians and other public health specialists to forge new relationships with environmental specialists and advocates. With the birth of the modern environmental movement in the late 1960s, public health and environmental specialists separated professionally as new departments of environmental protection were created, often out of old public health departments. It is time that these disciplines joined forces to work on the inseparable goals of preserving the environment and promoting the health and well-being of the global population.

References

1. Keyfitz, N. The growing human population. *Scientific American* 261 (1989), no. 3: 118–126.

2. Clark, W. Managing Planet Earth. *Scientific American* 261 (1989), no. 3: 47–54.

3. Brown, L. S., and Flavin, C. The earth's vital signs. In *The State of the World, 1988*. Norton.

4. *Our Common Future*. World Commission on Environment and Development, United Nations, 1987.

5. Case, B. W. Approaching environmental medicine. *Pennsylvania Medicine* 93 (1992): 52–55.

6. Levy, B. S. The teaching of occupational health in U.S. medical schools: Five-year follow-up of an initial survey. *American Journal of Public Health* 75 (1985): 79–80.

7. Gardner, M. J. Epidemiological studies of environmental exposure and specific diseases. *Archives of Environment Health* 43 (1990): 102–108.

8. Ris, C. H., and Preuss, P. V. Risk assessment and risk management: A process. In *Risk Assessment and Risk Management of Industrial and Environmental Chemicals,* ed. C. R. Gothern, M. A. Mehlman, and W. L. Marcus. Princeton Scientific, 1988.

9. Rall, D. P., et al. Alternatives to using human experience in assessing health risks. *Annual Review of Public Health* 8 (1987): 355–385.

10. National Research Council. *Improving Risk Communication.* National Academy Press, 1989.

11. McCallum, D., Morris, L., and Covello, V. The role of health professionals in environmental risk communications. Presented at Conference on Technological and Chemical Hazards: Community Awareness and Response, Washington, D.C., 1989.

12. Hendee, W. Stewardship of the environment. Presented at Conference on Technological and Chemical Hazards: Community Awareness and Response, Washington, D C , 1989

13. Goldstein, B. EPA as a public health agency. *Regulatory Toxicology and Pharmacology* 8 (1988): 328–334.

14. Case, B. W., and Mattison, D. R. Risk Assessment in Pathology Education. Manuscript, Center for Environmental Epidemiology, Graduate School of Public Health, University of Pittsburgh, 1992.

Urban and Transboundary Air Pollution: Human Health Consequences

David C. Christiani, M.D., M.P.H.

General Considerations

Recognition of the relationship between exposure to air pollutants and respiratory illness dates back to the 16th century, to vivid descriptions of respiratory disease in metalliferous miners.[1,2] In 1700 the epidemiologist and physician Ramazzini published the first textbook on occupational disease, *De Morbis Artificium Diatriba*,[3] in which he described in detail respiratory diseases due to exposure to silica, cotton, tobacco, flour, and other materials. In fact, much of what we currently understand about environmental lung disease derives from the study of exposed workers in the time since the industrial revolution.

Recognition of the relationship between non-workplace (i.e., community) air pollution and respiratory disease also dates back to the first use of coal as a combustion source, in the 14th century.[4] Later, in the industrial nations of Europe and North America, whole communities were acutely engulfed in air pollutants, resulting in serious illness and death among individuals with cardiopulmonary disease. These episodes occurred in the Meuse Valley of Belgium in 1930; in Donora, Pennsylvania, in 1948; and in London in 1952.[5] These air pollution emergencies were caused by air stagnation, which resulted in greatly increased concentrations of atmospheric pollutants, especially sulfur dioxide and suspended particulates. In the worst of these episodes, in London, the number of deaths ultimately reached over 4,000, a total second only to that of the influenza pandemic of 1917–18.[6] The observed marked excess in mortality was due mostly to bronchitis, pneumonia, and acute exacerbation of underlying cardiac and respiratory diseases.

As a result of these epidemics, there was increased attention by scientists and governments to the health effects of air pollution—identifying specific pollutant sources and their transport in the atmosphere, elucidating exposure-response relationships, and developing air pollution control technology. Despite these efforts, we face a crisis of worldwide air pollution today. This crisis has several aspects:

- Since the atmosphere is dynamic and always changing, contaminants are transported (sometimes over thousands of miles), diluted, precipitated, and transformed. Pollution, therefore, knows no boundaries or national borders.

- The primary emissions of sulfur oxides, nitrogen oxides, carbon monoxide, respirable particulates, and metals (e.g., lead and cadmium) are severely polluting cities and towns in Asia, Africa, Latin America, and Eastern Europe. Many cities in the Third World and in Eastern Europe are experiencing uncontrolled industrial expansion, growing motor vehicle numbers and congestion, and polluting cooking methods and heating fuels. A 1988 study of developing countries described the relationship between air pollution and wealth (GNP per capita): poorer countries (relying heavily on coal) had significantly higher levels of total suspended particulates than wealthier nations.[7] For example, in Beijing winter days are often dark and gray because suspended particulates exceed $500 \ \mu g/m^3$.[8] Similarly, coal-burning townships in South Africa have been found to have particulate concentrations exceeding $1,000 \ \mu g/m^3$.[9,10]

- In the nations that have reduced the primary emissions from heavy industry, power plants, and automobiles, new problems have arisen from pollution by newer industries and from air pollution caused by secondary formation of acids and ozone. The resulting high levels of air pollution (including indoor air pollution) are considered by experts to explain the high mortality rates for acute respiratory disease (about 4 million per year) in children under the age of 5 in many developing countries.[11]

Although this report is focused upon the human health effects of air pollution, there are also many other aspects of the problem. For example, damage to ecosystems and agriculture from acid rain, damage to buildings and artworks, and reduced visibility are all attributable to air pollution. Although industrialization has meant that a large number of pollutants have been released into the environment, this review will concentrate on what has been learned about the respiratory health effects of the most common urban air pollutants of the late 20th century.

Defining Adverse Health Effects

What constitutes an adverse human health effect has been a subject of scientific debate.[12] For the purpose of this review, any effect that results in altered structure or impaired function, or represents the beginning of a sequence of events leading to altered structure or function, will be considered an "adverse health effect."

Specific Air Pollutants Associated with Adverse Respiratory Effects

Several major types of air pollution are currently recognized to cause adverse respiratory health effects: sulfur oxides and acidic particulate complexes, photochemical oxidants, and a miscellaneous category of pollutants arising from industrial sources. Some of these commonly encountered pollutants are listed in table 1.

Sulfur Dioxide and Acidic Aerosols

Sulfur dioxide (SO_2) is produced by the combustion of sulfur contained in fossil fuels such as coal and crude oil. Therefore, the major sources of environmental pollution with sulfur dioxide are electric power generating plants, oil refineries, and smelters. Some fuels, such as "soft" coal, are particularly sulfur-rich. This has profound implications for nations such as China, which possesses 12% of the world's bituminous coal reserves and depends mainly on coal for electric power generation, steam, heating, and (in many regions) household cooking fuel.

Table 1 Principal air pollutants, their sources, and their respiratory effects.

Pollutant	Sources	Health effects
Sulfur oxides, particulates	Coal and oil power plants Oil refineries, smelters Kerosene stoves	Bronchoconstriction Chronic bronchitis Chronic obstructive lung disease
Carbon monoxide	Motor vehicle emissions Fossil fuel burning	Asphyxia leading to heart and nervous system damage, death
Oxides of nitrogen (NO_x)	Automobile emissions Fossil fuel power plants Oil refineries	Airway injury Pulmonary edema Impaired lung defenses
Ozone (O_3)	Automobile emissions Ozone generators Aircraft cabins	Same as NO_x
Polycyclic aromatic hydrocarbons	Diesel exhaust Cigarette smoke Stove smoke	Lung cancer
Radon	Natural	Lung cancer
Asbestos	Asbestos mines and mills Insulation Building materials	Mesothelioma Lung cancer Asbestosis
Arsenic	Copper smelters Cigarette smoke	Lung cancer
Allergens	Pollen Animal dander House dust	Asthma, rhinitis

Adapted, with permission, from H. A. Boushey and D. Sheppard, "Air pollution," in *Textbook of Respiratory Medicine,* ed. J. F. Murray and J. A. Nadel (Saunders, 1988).

Sulfur dioxide is a clear, highly water-soluble gas, so it is effectively absorbed by the mucous membranes of the upper airways, with a much smaller proportion reaching the distal regions of the lung.[13] The sulfur dioxide released into the atmosphere does not remain gaseous. It undergoes chemical reaction with water, metals, and other pollutants to form aerosols. Statutory regulations promulgated in the early 1970s by the U.S. Environmental Protection Agency under the Clean Air Act resulted in significant reductions in levels of SO_2 and particulates. However, local reductions in pollution were often achieved by the use of tall stacks, particularly for power plants. This resulted in the pollutants' being emitted high into the atmosphere, where prolonged residence time allowed their transformation into acid aerosols. These particulate aerosols vary in composition from area to area, but the most common pollutants resulting from this atmospheric reaction are sulfuric acid, metallic acids, and ammonium sulfates. Sulfur dioxide, therefore, together with other products of fossil fuel combustion (soot, fly ash, silicates, metallic oxides, etc.), forms the heavy urban pollution that typified old London, many Third World cities today which mainly burn coal, as well as basin regions in the United States such as in areas of Utah where there are coal-burning plants.

In addition to this smog, which is a descriptive term generically referring to the visibly cloudy combination of smoke and fog, an acidic aerosol is formed which has been shown to induce asthmatic responses in both adults and children.[14-16] The Harvard Six Cities study demonstrated a significant association between chronic cough and bronchitis and hydrogen ion concentration—a measure of acidity—rather than sulfate levels or total particulate levels.[17] Also, both epidemiologic and controlled human studies have demonstrated remarkable sensitivity in exercising asthmatics to the bronchoconstrictive effects of acidic aerosols.[18,19] Several studies have also linked exposure to acidic aerosols and mortality,[20,21] documenting an increase in deaths in persons with underlying chronic heart and lung disease who had been exposed. Finally, acidic aerosols result in "acid rain," which may threaten aquatic life.[22]

The acute bronchoconstrictive effects of sulfur dioxide have been well known for almost 30 years.[23,24] For example, inhalation of high concentrations of SO_2 has been shown to increase airway resistance in healthy normal volunteers[23,24]—profoundly, in asthmatics.[18,19,25] Although this effect in healthy humans is noted only after inhalation of high concentrations (e.g., 5 ppm), well above the concentrations found in most polluted cities, other human studies have shown that patients with asthma are exquisitely sensitive to the bronchoconstrictive effects of SO_2 and react to much lower levels.[25-28] Because SO_2 is highly water soluble, nearly all of the inspired SO_2 gas is removed in the upper airways during rest. Exertion (work or exercising) will increase the fraction of SO_2 gas to the lower airways and therefore help to precipitate bronchoconstriction. It has been demonstrated that concentrations of

Table 2   U.S. National Air Quality Standards. Source: reference 49.

Pollutant	Maximum allowable concentration	Averaging period
Sulfur dioxide	0.14 ppm	24 hours
	0.03 ppm	1 year
Nitrogen dioxide	0.05 ppm	1 year
Ozone	0.12 ppm	1 hour
Carbon monoxide	35 ppm	1 hour
	9 ppm	8 hours
Lead	1.5 $\mu g/m^3$	1 month
Total suspended particulates	260 $\mu g/m^3$	24 hours
	75 $\mu g/m^3$	1 year

SO_2 as low as 0.4 ppm can cause symptomatic bronchoconstriction when inhaled by exercising asthmatics. These levels may be encountered in polluted urban air, and exposure to these levels for even several minutes is enough to induce bronchospasm.[29] So, sulfur-dioxide-induced bronchoconstriction is not necessarily prevented by the current EPA standard of 0.14 ppm as a 24-hour maximal average, as this standard does not set limits for maximal concentrations over shorter periods (table 2).

In addition to the acute bronchoconstrictive effects of sulfur dioxide, there is epidemiologic evidence for chronic airway obstruction in persons exposed to elevated levels of SO_2.[30,31]

Particulates

Particulate air pollution is closely related to SO_2 and aerosols. The term usually refers to particles suspended in the air after various forms of combustion or other industrial activity. In the epidemics noted in the introduction, the air pollution was characterized by high levels of particulates, sulfur dioxide, and moisture. Recent studies have shown that particulate air pollution *per se* was associated with increases in daily mortality in London in both the heavy smog episodes of the 1950s and the lower pollution levels of the late 1960s and the early 1970s.[32,33] A study performed in Steubenville, Ohio, between 1974 and 1984 revealed that total suspended particulate counts were significantly asso-

---

Table 3 Indoor air pollution in developing countries. Reproduced, with permission, from reference 35.

	Conditions	Particulate concentration ($\mu g/m^3$)
New Guinea	Overnight	200–9,000
Kenya	Highlands	2,700–7,900
	Lowlands	300–1,500
	24 hours	1,200–1,900
India	Cooking material	
	wood	15,000
	dung	18,300
	charcoal	5,500
	General	4,000–21,000
Nepal	Cooking (wood)	4,700
China	Cooking (wood)	2,600
Gambia	24 hours	1,000–2,500
During cooking		
India	Villages	3,600–6,800
Nepal	Villages*	8,200
	Villages†	3,000

*before introduction of new cooking stoves
†after introduction of new cooking stoves

ciated with increased daily mortality, with season and temperature controlled for.[34] An increase in particulates of 100 ng/m^3 was associated with a 40% increase in mortality on the succeeding day. The striking quantitative agreement in the relative increase in mortality among studies done in London, New York, and Steubenville further strengthened the evidence for a causal association.

The very high particulate levels (both indoor and outdoor) measured in developing countries (table 3) range up to 100 times the current U.S. standard of 150 $\mu g/m^3$ for particles less than 10 μm in diameter.[35] The health consequences, as one would expect from the above discussion, are great. Deaths in children from acute respiratory disease are high, even in nations with lower overall infant mortality.[11] For example, Western Europe has a low overall infant mortality rate, yet has 3 times North America's rate of infant respiratory-disease mortality.[11] The rate in Eastern Europe is nearly 20 times that in North America. After infant diarrhea, acute respiratory disease is believed to be the major cause of

CO concentrations dropped from approximately 10 ppm in 1970 to 3 ppm in the 1980s.[50] However, because of increased vehicle use and traffic congestion, there remain many localized "hot spots" in U.S. cities where CO levels still exceed air-quality standards. Exposure can also occur after the burning of coal, paper, oil, gas, or any other carbonaceous material.

The health effects of carbon monoxide have been documented in clinical observations of patients with CO intoxication and in experimental and epidemiologic studies of persons exposed to low-level CO. Carbon monoxide is an odorless, colorless gas produced by incomplete combustion of carbonaceous fuels such as wood, gasoline, and natural gas.[51] Because of its marked affinity for hemoglobin, CO impairs oxygen transport, and poisoning often manifests as adverse effects in the cardiovascular and central nervous systems, with the severity of the poisoning directly proportional to the level and duration of the exposure. Thousands of people die annually (at work and at home) from CO poisoning, and an even larger number suffer permanent damage to the central nervous system. A sizable portion of the workers in any country have significant CO exposure, as do a larger proportion of persons living in poorly ventilated homes where biofuels are burned.

The pathophysiology of CO poisoning can be conceptualized as anti-oxygen activity, since CO is an antimetabolite of oxygen. Inhaled CO binds strongly to hemoglobin in the pulmonary capillaries, resulting in the complex called carboxyhemoglobin (COHb). The affinity of human hemoglobin for CO is about 240 times its affinity for oxygen.[52] The formation of COHb has two considerable effects: it blocks oxygen carriage by inactivating hemoglobin, and its presence in the blood shifts the dissociation curve of oxyhemoglobin to the left so that the release of remaining oxygen to tissues is impaired. Because of this latter effect, the presence of any level of COHb in the blood, above the background from catabolism of 0.3–0.8%, will interfere with tissue oxygenation much more than an equivalent reduction in hemoglobin from anemia or bleeding. Carbon monoxide also binds with myoglobin to form carboxymyoglobin, which may disturb muscle metabolism (especially in the heart).

The clinical effects of acute CO poisoning vary with the level of COHb, ranging from nonspecific symptoms (headache, dizziness, fatigue) to death.[53] Generally, acute symptoms appear at COHb levels of

10% or greater, and levels above 50% are associated with collapse, convulsions, and death. Because of the profound physiological effects associated with increasing levels of COHb, individuals with certain underlying conditions are at particularly high risk for CO poisoning. These conditions include chronic hypoxemia from lung or heart disease, cerebrovascular disease, peripheral vascular disease, arteriosclerotic heart disease, anemia, and hemoglobinopathies.[53] Fetuses and children are more susceptible than adults to CO poisoning.[54,55]

Chronic, lower-level exposure to CO has also been postulated to accelerate atherosclerotic vascular disease by affecting cholesterol uptake in the arterial wall, though results from animal and *in vitro* studies are conflicting.[56-59] There is also some evidence that CO may accelerate clot lysis time[60] and may increase platelet activity and coagulation.[61] These alterations in the clotting system could increase the risk for thromboembolism in the heart or the brain.

In addition to sulfur dioxide, particulates, and photochemical pollutants, urban air contains a number of known carcinogens, including polycyclic aromatic hydrocarbons (PAHs), n-nitroso compounds, and, in many regions, arsenic and asbestos. Exposure to these compounds is associated with increased risk of lung cancer in various occupationally exposed groups (e.g., coke oven workers exposed to PAHs and insulators exposed to asbestos). Therefore, populations living near coke ovens or exposed to asbestos insulation at home and in public buildings also may be at increased risk of lung cancer.

In addition to these agents, airborne exposure to the products of waste incineration, such as dioxins and furans, is on the increase in some nations. Though usually present in communities in concentrations much lower than those found in workplaces, airborne dioxins, furans, and other incineration products may still lead to increased lung-cancer risks, particularly in neighborhoods or villages near point sources where their levels may be substantial. The degree of cancer risk for ambient exposures to these compounds has not been calculated to date.

Indoor Air Pollution

No description of the health effects of air pollution is complete without a discussion of indoor air pollution. The descriptions above generally have referred to community ("ambient") air pollution. Moreover, as mentioned above, much of what we now know about the health effects of airborne toxins derives from the study of workers in factories and mines—both of which are "indoor" environments. However, today, both the common and the scientific use of the term "indoor air pollution" refer to homes and non-factory public buildings (modern office buildings, hospitals, etc.). Most indoor environments, whether they be traditional village homes in the Third World or tightly sealed high-rise office buildings, have air pollutant sources. Pollution can come from heating and cooking combustion, pesticides, tobacco emissions, abrasion of surfaces, evaporation of vapors and gases, radon, and microbiologic material from people and animals.[62] High concentrations of pollution in indoor settings in either the developed world or the developing world can be associated with mucous-membrane irritation, discomfort, illness, and even death.[62]

Although indoor air pollution has increased substantially in the industrialized nations because of tighter building construction and because of the widespread use of building materials that may give off gaseous chemicals, indoor air pollution is a particular problem in the poor communities of developing countries. Wood, crop residues, animal dung, and other forms of biomass are used by approximately half the world's population (2.5 billion people) as cooking and/or heating fuels, often in poorly ventilated conditions. This leads to high exposures to such air pollutants as carbon monoxide and polycyclic aromatic hydrocarbons, particularly for women and children. The health effects of such exposures are summarized in table 1.

There are strong indications that indoor air pollution is associated with acute respiratory infection in infants and in children under 5 years. In view of the large numbers of children from developing countries who die from respiratory infection each year, this is a high priority for public health research and action.[63]

Current Regulation of Air Pollution

Approaches to controlling air pollution have varied from nation to nation. For example, in Great Britain the approach was to recognize that outdoor urban air—not specified as to pollutants—was unhealthy, and the government limited emission of a mixture of suspected pollutants with its Clean Air Act of 1956.

The approach in the United States has been different. Congress passed a Clean Air Act in 1970 which required the Environmental Protection Agency to define air quality standards "allowing a margin of safety" to protect the public's health. This required identifying specific pollutants and setting standards for each. The pollutants chosen as criteria were sulfur dioxide, carbon monoxide, nitrogen dioxide, ozone, lead, and total suspended particulates. The assumption underlying this approach is that scientific research would allow the identification of a threshold concentration for each pollutant, below which adverse health effects would not occur. Even with the inherent problems of this assumption, standards were set for all six pollutants (see table 2). Therefore, the federal government prescribed procedures and standards for the states to follow. The laws dictate that polluters file permits and maintain documentation, and that failure to comply will result in potential fines, loss of operating permits, liability for cleanup, and civil or even criminal penalties for cleanup and/or injury. There are obvious instances where compliance with the law still does not ensure health protection against air pollutants, or ensure environmental protection. For example, in the construction of new highways, the environmental impact review process assesses compliance with air quality standards at outdoor locations. Ignored in the assessment for the highways are other important impacts, such as the associated industrial and construction projects, the increase in miles traveled in vehicles (and the consequent emissions), and the increased general energy expenditure for utilities such as electricity.

The limitations of the current U.S. approach to air pollution can be briefly summarized as follows:

- Ozone and lead are clearly not adequately controlled under the current standards (section 109 of the Clean Air Act). Also, there is inadequate control of acidity inherent in the SO_2 limits. Moreover, in section 112 (which addresses carcinogens), the U.S. government's approach has been not to force new technological development of controls, but rather to use the least expensive available methods.
- There has been a failure to address indoor air pollutants.
- There has been a failure to adequately address non-criteria airborne pollutants that affect human health through indirect mechanisms—

---

i.e., CO_2 and chlorofluorocarbon emissions as causes of global warming and stratospheric ozone depletion, respectively.

The problem of controlling air pollution is indeed a pressing one. Atmospheric pollution has now reached a level that threatens not only the health of entire populations but also their survival. Various national regulatory approaches have not, so far, been up to the task of controlling pollution on a global scale. Air pollution is a growing, global problem. Only global approaches will succeed in controlling it.

References

1. Agricola, G. *De Re Metallica,* Basel, 1556. Translation in *Mining Magazine* (London, 1912).

2. Paracelsus, T. *Von der Bergsacht.* Dilingen, 1567.

3. Ramazzini, B. *De Morbis Artificium Diatribe.* Translation: *Disease of Workers* (University of Chicago Press, 1940).

4. Ayres, S. M., Evans, R. G., and Buehler, M. E. Air pollution: A major public health problem. *CRC Critical Review of Laboratory Science* 3 (1972): 1–40.

5. Shy, C. M., et al. *Health Effects of Air Pollution.* New York Lung Association, 1978.

6. Logan, W. P. D. Mortality in the London fog incident. *Lancet* 1 (1953): 336–339.

7. Smith, K. R. Air pollution: Assessing total exposure in developing countries. *Environment* 30 (1988): 16.

8. Urban Air Pollution in the People's Republic of China, 1981–84. World Health Organization report EHE-EFP/85-5.

9. Kemeny, E., Ellerbeck, R. H., and Briggs, A. B. An assessment of smoke pollution in Soweto: Residential air pollution. In *Proceedings of the National Association for Clean Air, 10–11 Nov., 1988,* Pretoria, South Africa.

10. Tshangwe, H. T., Kgamphe, J. S., and Annegarn, H. J. Indoor-outdoor air particulate study in a Soweto home: Residential air pollution. In *Proceedings of the National Association for Clean Air, 10–11 Nov., 1988,* Pretoria, South Africa.

11. Leowski, J. Mortality from acute respiratory infections in children under 5 years of age; global estimates. *World Health Statistics Quarterly* 39 (1986): 138–144.

12. Andrews, C., et al. Guidelines as to what constitutes an adverse respiratory health effect, with special reference to epidemiologic studies of air pollution. *American Review of Respiratory Diseases* 131 (1985): 666–668.

13. Frank, N. R., et al. SO_2 (^{35}S-labeled) absorption by the nose and mouth under conditions of varying concentration and flow. *Archives of Environmental Health* 18 (1969): 315–322.

14. Bates, D. V., and Sizto, R. Air pollution and hospital admissions in Southern Ontario: The acid summer haze effect. *Environmental Research* 43 (1987): 317–331.

15. Bates, D. V., and Sizto, R. The Ontario air pollution study: Identification of the causative agent. *Environmental Health Perspectives* 79 (1989): 69–72.

16. Ostro, B. D., et al. Asthmatic responses to airborne acidic aerosols. *American Journal of Public Health* 81 (1991): 694–702.

17. Ware, J. H., et al. Effects of ambient sulfur dioxides and suspended particles on respiratory health of preadolescent children. *American Review of Respiratory Diseases* 133 (1986): 834–842.

18. Bauer, M. A., et al. Inhalation of 0.30 ppm nitrogen dioxide potentiates exercise-induced bronchospasm in asthmatics. *American Review of Respiratory Diseases* 134 (1986): 1203–1208.

19. Koenig, J. Q., Covert, D. S., and Pierson, W. E. Effects of inhalation of acidic compounds on pulmonary function in allergic adolescent subjects. *Environmental Health Perspectives* 79 (1989): 173–178.

20. Thurston, G. D., et al. Reexamination of London, England, mortality in relation to exposure to acidic aerosols during 1963–71 winters. *Environmental Health Perspectives* 79 (1989): 73–82.

21. Ostro, B. D. A search for a threshold in the relationship of air pollution to mortality: A reanalysis of data on London winters. *Environmental Health Perspectives* 58 (1984): 397–399.

22. German, E., Martin, F., and Litzau, J. T. Acid rain: Ionic correlations in the Eastern United States, 1980–1981. *Science* 225 (1984): 407–409.

23. Frank, N. R., et al. Effects of acute controlled exposure to SO_2 on respiratory mechanics in healthy male adults. *Journal of Applied Physiology* 17 (1962): 252–258.

24. Nadel, J. A., et al. Mechanisms of bronchoconstriction during inhalation of sulfur dioxide. *Journal of Applied Physiology* 20 (1965): 164–167.

25. Sheppard, D., et al. Lower threshold and greater bronchomotor responsiveness of asthmatic subjects to sulfur dioxide. *American Review of Respiratory Diseases* 122 (1980): 873–878.

26. Sheppard, D., Saisho, A., and Nadel, J. A. Exercise increases sulfur dioxide-induced bronchoconstriction in asthmatic subjects. *American Review of Respiratory Diseases* 123 (1981): 486–491.

27. Koenig, J. Q., et al. Effects of SO_2 plus NaCl aerosol combined with moderate exercise on pulmonary function in asthmatic adolescents. *Environmental Research* 25 (1981): 340–348.

28. Linn, W. S., et al. Respiratory effects of sulfur dioxide in heavily exercising asthmatics: A dose-response study. *American Review of Respiratory Diseases* 127 (1983): 278–283.

29. Sheppard, D., et al. Magnitude of the interaction between the bronchomotor effects of sulfur dioxide and those of dry (cold) air. *American Review of Respiratory Diseases* 130 (1984): 52–55.

30. Xu, X. P., Dockery, D. W., and Wang, L. H. Effects of air pollution on adult pulmonary function. *Archives of Environmental Health* 46 (1991): 198–206.

31. Holland, W. W., and Reid, D. D. The urban factor in chronic bronchitis. *Lancet* 1 (1965): 445–448.

32. Mazumdar, S., Schimmel, H., and Higgins, I. T. T. Relation of daily mortality to air pollution: an analysis of 14 London, England winters, 1958/59–1971/72. *Archives of Environmental Health* 37 (1982): 213–220.

33. Schwartz, J., and Marcus, A. Mortality and air pollution in London: A time-series analysis. *American Journal of Epidemiology* 131 (1990): 185–194.

34. Schwartz, J., and Dockery, D. W. Particulate air pollution and daily mortality in Steubenville, Ohio. *American Journal of Epidemiology* 135 (1992): 12–19.

35. Pandey, M. R., et al. Indoor air pollution in developing countries and acute respiratory infection in children. *Lancet* 1 (1989): 427–429.

36. Photochemical Oxidants—Environmental Health Criteria 7. World Health Organization report CIS 80-1919, 1979.

37. Niki, H., Dahy, E. E., and Weinstock, B. Mechanisms of smog reactions. *Advances in Chemistry* 113 (1972): 16–57.

38. Miller, J. F. Similarity between man and laboratory animals in regional pulmonary deposition of ozone. *Environmental Research* 17 (1978): 84–101.

39. Bates, D. V., et al. Short-term effects of ozone on the lung. *Journal of Applied Physiology* 32 (1972): 176–181.

40. Hazucha, M., et al. Pulmonary function in man after short-term exposure to ozone. *Archives of Environmental Health* 27 (1973): 183–188.

41. Hackney, J. D., et al. Experimental studies on human health effects of air pollutants. *Archives of Environmental Health* 30 (1975): 373–378.

42. Hackney, J. D., et al. Adaptation to short-term respiratory effects of ozone in men repeatedly exposed. *Journal of Applied Physiology* 43 (1977): 82–85.

43. Graham, D. E., and Koren, H. S. Biomarkers of inflammation in ozone-exposed subjects. *American Review of Respiratory Diseases* 142 (1990): 152–156.

44. Coffin, D. L., and Stockinger, H. E. Biological effects of air pollutants. In *Air Pollution,* third edition, volume 2, ed. A. C. Stern. Academic Press.

45. Dungworth, D., Goldstein, E., and Ricci, P. F. Photochemical air pollution—Part II. *Western Journal of Medicine* 142 (1985): 523–531.

46. Hackney, J. D., et al. Experimental studies on human health effects of air pollutants: IV. Short-term physiological and clinical effects of nitrogen dioxide exposure. *Archives of Environmental Health* 33 (1978): 176–181.

47. Golden, J., Nadel, J. A., and Boushey, H. A. Bronchial hyperirritability in healthy subjects after exposure to ozone. *American Review of Respiratory Diseases* 118 (1978): 287–294.

48. Holtzman, M. J., et al. Effect of ozone on bronchial reactivity in atopic subjects. *American Review of Respiratory Diseases* 122 (1980): 17–25.

49. National Air Quality and Emission Trends Report, 1987. EPA publication EPA/450/4-89/001.

50. Kuntasal, G., and Chang, T. Y. Trends and relationships of ozone, NO_x and HC in the south coast air basin of California. *Journal of the Air Pollution Control Association* 35 (1987): 1158–1163.

51. National Research Council Committee on Medical and Biological Effects of Environmental Pollutants. *Carbon Monoxide*. National Academy of Sciences, 1977.

52. Coburn, R. F., Forster, R. E., and Kane, P. B. Considerations of the physiological variables that determine blood carboxyhemoglobin concentration in man. *Journal of Clinical Investigation* 44 (1965): 1899–1910.

53. Winter, P. M., and Miller, J. N. Carbon monoxide poisoning. *Journal of the American Medical Association* 236 (1976): 1502–1504.

54. Longo, L. D. The biological effects of carbon monoxide on the pregnant woman, fetus, and newborn infant. *American Journal of Obstetrics and Gynecology* 129 (1977): 69–103.

55. Zimmerman, S. S., and Truxal, B. Carbon monoxide poisoning. *Pediatrics* 68 (1981): 215–224.

56. Astrup, P., Kjeldsen, J., and Wanstrup, J. Effects of carbon monoxide exposure on arterial walls. *Annals of the New York Academy of Sciences* 174 (1970): 294–300.

57. Theodore, J., O'Donnell, R. D., and Back, K. C. Toxicological evaluation of carbon monoxide in humans and other mammalian species. *Journal of Occupational Medicine* 13 (1971): 242–255.

58. Sarma, S. J. M., et al. The effect of carbon monoxide on lipid metabolism of human coronary arteries. *Atherosclerosis* 22 (1975): 193–195.

59. Armitage, A. K., Davies, R. F., and Turner, D. M. The effects of carbon monoxide on the development of atherosclerosis in the white carneau pigeon. *Atherosclerosis* 23 (1976): 333–344.

60. El-Attar, O. A., and Suiro, D. M. Effect of carbon monoxide on the whole fibrinolytic activity. *Industrial Medicine and Surgery* 37 (1968): 774–777.

61. Haft, J. I. Role of blood platelets in coronary artery disease. *American Journal of Cardiology* 43 (1979): 1197–1206.

62. Samet, J. M., Marbury, M. C., and Spengler, J. D. Health effects and sources of indoor air pollution. Part I. *American Review of Respiratory Diseases* 137 (1988): 221–242.

63. *Epidemiological, Social and Technical Aspects of Indoor Air Pollution from Biomass Fuel*. World Health Organization, 1991.

Drinking-Water Pollution and Human Health

Howard Hu, M.D., M.P.H., Sc.D.
Nancy K. Kim, Ph.D.

The quest for pure drinking water in the face of contamination has existed since ancient times. John Snow, in 1855, was the first to prove that public water supplies could be a source of infection for humans.[1] It was not until the 1960s and 1970s, however, that the public health community began to recognize that hazardous substances related to industrialization and agricultural activity were a major threat to the safety of water used for drinking and bathing through contamination of surface water and ground water. In the United States, several federal environmental protection statutes were enacted to curtail discharges to groundwater and surface water, and in 1974 Congress passed the Safe Drinking Water Act (which requires the setting of standards for all possible pollutants in drinking water).[1] Similar legislation has been passed in other countries.

Despite these measures, the threat to health posed by water contamination remains high in the United States and elsewhere. Nearly half the population of the United States use groundwater from wells or springs as their primary source of drinking water,[2] and groundwater and surface water remain the world's main source of drinking water. Contamination can be difficult to detect, and extremely difficult to reverse once detected. Furthermore, many inorganic and organic chemicals that had been deemed safe are now thought to be hazardous to human health, as investigations into their toxicity have progressed.

These problems are gradually spreading to developing countries, too, beginning with contamination in urban environments and spreading to rural communities. A recent review of case examples from Nigeria, Kenya, Brazil, and India demonstrated the complexity of these problems and the need for greater expertise and a more integrated multi-objective approach to water-resource management.[3]

Of the four (physical, chemical, biologic, and radioactive) general characteristics of water quality,[4] we shall concentrate on chemical and radioactive pollutants. Most physical aspects of water quality (taste, odor, color, temperature, turbidity, suspended solids, and mineral content) are not of primary concern with respect to human health (although it should be noted that water with a mineral content in excess of 500 milligrams per liter can have a laxative effect). Improved recognition and prevention of contamination by biological agents is of obvious continuing concern, particularly in the developing world, but is well reviewed elsewhere.[5,6]

Sources of Contamination and Routes of Exposure

Surface water can be contaminated by point or non-point sources. A runoff pipe from an industrial plant or a sewage-treatment plant discharging chemicals into a river is a point source; the carrying of pesticides by rainwater from a field into a lake is a non-point discharge.[7] Fresh surface water can also be affected by groundwater quality; for example, approximately 30% of the streamflow of the United States is supplied by groundwater emerging as natural springs or other seepage.[8]

Groundwater is contained in a geological layer termed an *aquifer*. Aquifers are composed of permeable or porous geological material, and may either be unconfined (and thereby most susceptible to contamination) or confined by relatively impermeable material called *aquitards* (see figure 1). Though they are located at greater depths and are protected to a degree, confined aquifers can nevertheless be contaminated when they are tapped for use or are in proximity for a prolonged period of time to a source of heavy contamination.

Contamination of aquifers can occur via the leaching of chemicals from the soil, from industrial discharges, from septic tanks, or from underground storage tanks. Fertilizers applied to agricultural lands con-

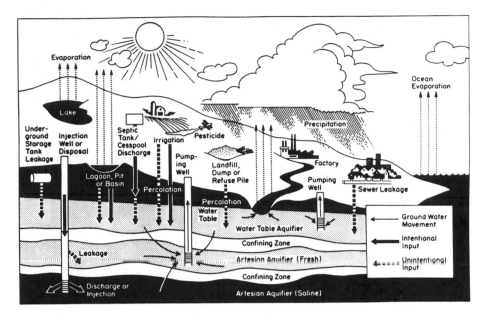

Figure 1   How waste-disposal practices can contaminate the groundwater system.
Source: reference 2.

tain nitrates which dissolve easily in water. Rainwater percolating
through the soil can carry the dissolved nitrates with it into aquifers.
Industry or homes can discharge wastewater directly into groundwater
from septic tanks or waste holding tanks. Buried underground tanks
used to store chemicals, such as gasoline or fuel oil, can leak, allowing
their contents to seep into groundwater.

The chemical characteristics of a contaminant may change as it
percolates through the soil zone to the aquifer. Attenuation may occur
through a number of processes, such as dilution, volatilization, mechanical filtration, precipitation, buffering, neutralization, microbial metabolism, and plant uptake. These generally reduce the toxicity of the
contaminant. Once a contaminant gains entry into an aquifer, transport
usually results in an elliptical plume of contamination, the shape, flow
rate, and dispersion of which depend on aquifer permeability, hydraulic
gradients, contaminant chemistry, and many other factors.

People can be exposed to polluted groundwater or surface water
through a number of routes. Most commonly, contaminated water can
be collected or pumped and used directly for drinking or cooking.

Significant exposure to chemicals in surface water can also occur when swimming in a lake or river. Some chemicals accumulate in fish that are subsequently caught and eaten. A chemical that volatilizes easily can escape from groundwater and rise through soil, and in gaseous form the chemical can then be released into surroundings or can enter homes through cracks in basements, exposing residents through inhalation. If water used for bathing is contaminated, some chemicals can also be absorbed through skin or inhaled in the fine spray of a shower. Of these routes of exposure, use of contaminated water for drinking and cooking is clearly the most dominant threat, followed by ingestion of contaminated fish (especially in areas where high fish consumption and pollution coexist).[9]

Most of the contaminants in surface water and groundwater that are due to human activity derive from agricultural and industrial sources. The spectrum of contaminants is enormous. The most important ones are toxic heavy metals (such as lead, arsenic, cadmium, and mercury), pesticides and other agricultural chemicals (such as nitrates, chlorinated organic chemicals (DDT), organophosphate or carbamate (aldicarb) insecticides, and herbicides (2,4-D), and volatile organic chemicals (such as gasoline products and the halogenated solvents trichloroethene and tetrachloroethene). There are also some natural sources of hazardous chemical exposure; for example, deep wells are often contaminated with naturally occurring arsenic.

Specific Hazards

Nitrates and Nitrites

From a global perspective, biological processes such as nitrogen fixation and the conversion of organic nitrogen to ammonia (NH_3) or nitrate (NO_3) are the major sources of inorganic nitrogen compounds in the environment. However, on a local scale, municipal and industrial wastewaters (particularly sewage treatment plants, fertilizers, refuse dumps, septic tanks, and other sources of organic waste) are major nitrogen sources.[2] Waste sources are significant, can greatly exceed natural sources, and are increasingly found in groundwater primarily because of a marked rise in the use of nitrogenous fertilizers around the world.

Most human intake of nitrogen normally comes from food rather than water. Vegetables are the highest source of nitrates for most people. Of lesser importance is the contribution of nitrates from cured meats.[7] Cured meats, baked goods, and cereals contribute most of the dietary nitrite; vegetables provide much less.[10]

The total nitrogen content of water is usually measured for both nitrates and nitrites. Nitrate or nitrite is more likely to be found in higher concentrations in groundwater than in surface water, and shallow wells (especially dug wells) are more likely to be contaminated than deep or drilled wells.[2] The drinking-water standard for nitrate in the United States is 10 mg per liter of nitrate (measured as total nitrogen).[11] This level is often exceeded; for example, in wellwater surveys in South Dakota, up to 39% of dug or bored wells were found to contain nitrates above this level.[12]

Two potential health effects of concern from nitrate or nitrite in drinking water are the induction of methemoglobinemia and the formation of nitrosamines. Nitrate itself is relatively non-toxic to humans; however, when converted in the body to nitrite (primarily by bacteria in the colon) and absorbed, this nitrogen compound is capable of oxidizing hemoglobin (the principal oxygen transport molecule of the body), with the consequent induction of methemoglobinemia and oxygen starvation.[8] The symptoms of methemoglobinemia include bluish skin and lips, weakness, rapid pulse, and tachypnea. Infants are particularly at risk, because the higher pH of an infant's stomach contents provides a more favorable environment for the bacteria that convert nitrate to nitrite. Babies less than 1 year of age[13] and babies with respiratory illnesses or diarrhea[14] may be at greater risk from nitrite-induced methemoglobinemia.

The risk for methemoglobinemia from drinking water containing less than 10 mg of nitrate (as nitrogen) per liter is relatively low. With drinking water contaminated by nitrates at levels above 10 mg/l, however, the risk is significant; around 17–20% of infants develop methemoglobinemia when exposed to these higher levels.

It has been postulated that nitrates can also form nitrosating agents, which can react with secondary organic amines to form nitrosamines. Nitroso compounds are carcinogenic at high doses in animal studies.[7]

The formation of nitrosamines from nitrates, however, is primarily theoretical. Examination of mortality data for 253 urban areas in the United Kingdom in relation to nitrate levels in water showed no evidence of a positive association between nitrate levels and the risk of stomach cancer.[15] To date, nitrates in drinking water have not convincingly been shown to cause cancer.[16]

Some studies have found increases in the risk of malformations in children of women consuming water with concentrations of nitrate greater than 5 mg/l;[17,18] however, it is not clear whether this link is causal. Fan et al.[19] reviewed the literature and did not find enough evidence to conclude that nitrate exposure produces teratogenic or adverse reproductive effects. They also suggested that a 10-mg/l nitrate standard for drinking water would be protective against these effects.

Heavy Metals

The heavy metals of greatest concern for health with regard to environmental exposure through drinking water are lead and arsenic. Cadmium, mercury, and other metals are also of concern, although exposure to them tends to be more sporadic. Significant levels of these metals may arise in drinking water, directly or indirectly, from human activity. Of most importance is seepage into groundwater of the run-offs from mining, milling, and smelting operations, which concentrate metals in ores from the earth's crust, and effluents and hazardous wastes from industries that use metals.[1] Lead contamination in drinking water is of particular concern, as lead was used in household plumbing and in the solder used to connect it. Seepage of heavy metals (especially arsenic) from the earth's crust can be a natural source of contamination in some areas where deep wells are used for drinking water. Quantitative modeling of the movement of heavy metals in the environment suggests that accumulation of metals in water, in soil, and in the food chain is accelerating around the world.[20]

In the ensuing brief discussion, emphasis will be placed on the sources of heavy-metal contamination of drinking water. A fuller discussion of health effects may be found in the next chapter.

Lead

The sources of lead in drinking water that are of greatest concern are lead pipes, the use of which was highly prevalent until the 1940s, and

---

lead solder, which was used (and is still being used in some countries) to connect plumbing. Also of concern is the seepage of lead from soil contaminated with the fallout from combusted leaded gasoline and the potential seepage of lead in hazardous-waste sites. In the United States, lead is prevalent in over 43% of hazardous-waste sites, and migration and groundwater contamination have been documented in almost half of these lead-containing sites.[21]

Lead contamination of drinking water from lead pipes and solder is more likely to be found in water samples taken at the tap than at the treatment plant. Soft water leaches more lead than does hard water, and the greatest exposure is likely to occur when a tap is first turned on after not being used for six or more hours.[22] (The water drawn in such a case is called *first-draw water.*)

Lake and river water, worldwide, contains about 1–10 micrograms of lead per liter. Because of lead in plumbing systems, lead levels in drinking water at the tap as high as 1,500 µg/l have been found. Drinking water is only one of many potential sources of lead exposure; lead paint, dust, food, and air pollution are other important sources, particularly in old urban areas.[1]

The current U.S. Centers for Disease Control level of concern for blood lead in children in 10 µg per deciliter.[23]. This level was set because of recent evidence linking low-level lead exposure to adverse effects on neurobehavioral development and school performance in children.[24] That concentration can generally be reached when a child ingests 8 ounces of contaminated water per day (defined as tap water with "first-draw water" containing more than 100 µg of lead per liter). In order to create a margin of safety, the Environmental Protection Agency recently lowered the amount of lead that is acceptable in drinking water from 50 to 15 µg/l. The importance of reducing children's exposure to lead is underscored by new evidence that suggests that cognitive deficits caused by lead are at least partly reversible.[56]

Arsenic
Drinking water is at risk for contamination by arsenic from a number of human activities, including the leaching of inorganic arsenic compounds used in pesticide sprays, the contamination of surface water by fallout from the combustion of arsenic-containing fossil fuel, and the leaching of mine tailings and smelter runoff. For example, in Perham,

Minnesota, groundwater contamination by an arsenic-containing grass-hopper bait led to wellwater arsenic concentrations of 11–21 mg/l and to documented illness.[25]

With chronic exposure at high levels, children are particularly at risk; the primary symptoms are abnormal skin pigmentation, hyper-keratosis, chronic nasal congestion, abdominal pain, and various cardio-vascular manifestations.[26] Some of these same problems were noticed in Taiwan by W. P. Tseng, who also documented "black foot" disease (a vasospastic condition thought to be caused by chronic arsenic exposure leading to gangrene of the extremities) and high rates of skin cancer.[27]

At lower levels of exposure, cancer is the outcome of primary concern. Occupational and population studies have linked chronic high-dose arsenic exposure to cancer of the skin, the lungs, the lymph glands, the bone marrow, the bladder, the kidneys, the prostate, and the liver.[28–30] Using a linear dose-response model to extrapolate risk, and imposing that risk on a large population, one would predict that significant numbers of people with chronic low-dose arsenic exposure would develop cancer. Understanding the true risk from low-level arsenic exposure is an area of active epidemiological research.

Other heavy metals
Contamination of water with other heavy metals has caused problems in isolated instances. In 1977, the National Academy of Science ranked the relative contributions of these metals in water supplies as a function of man's activities as follows.

very great: cadmium, chromium, copper, mercury, lead, zinc
high: silver, barium, tin
moderate: beryllium, cobalt, manganese, nickel, vanadium
low: magnesium.[1]

Of these metals, mercury and cadmium are probably the most toxic at the levels found in water.

High levels of environmental exposure to mercury occurs primarily through the consumption of food tainted by organic (and sometimes inorganic) mercury (see next chapter). However, the uses of mercury compounds that give rise to these exposures, such as the treatment of seeds with phenyl mercury acetate (used for its antifungal properties), can also lead, through runoff, to the contamination of surface water and groundwater. Similarly, short–chain alkyl mercury compounds are lipid-soluble and volatile; therefore they pose a risk of skin absorption and inhalation from bathing in contaminated waters. The most common symptoms of high-level organic mercury poisoning are mental disturbances, ataxia (loss of balance), gait impairment, disturbances of speech, constriction of visual fields, and disturbances of chewing and swallowing.[31] The toxicological implications of low-level mercury exposure are poorly understood.

Environmental exposure to cadmium has been increasing as a result of mining, refining, smelting, and the use of cadmium in industries such as battery manufacturing. Environmental exposure to cadmium has been responsible for significant episodes of poisoning through incorporation into foodstuffs; however, the same sources of cadmium for these overt episodes of poisoning, such as the use of cadmium-contaminated sewage sludge as fertilizer, can potentially cause contamination of ground and surface water used for drinking and bathing. High cadmium consumption causes nausea, vomiting, abdominal cramping, diarrhea, kidney disease, and increased calcium excretion (which leads to skeletal weakening). As in the case of mercury, the toxic effects of chronic exposure to low levels of cadmium are poorly understood. Recent studies have demonstrated an increased rate of mortality from cerebrovascular disease (e.g., stroke) in populations from cadmium-polluted areas.[32] One study has also indicated an association between cadmium levels in drinking water and prostatic cancer.[33]

Pesticides

In today's world, especially in developing countries, the use of pesticides has become inextricably linked with agricultural production. Included under the rubric of "pesticides" are insecticides, herbicides, nematicides, fungicides, and other chemicals used to attract, repel, or control pest growth. Insecticides and nematicides, including the bicyclophosphates, cyclodienes, and the pyrethroids, generally work by inhibiting neurotransmitter function in the peripheral and central nervous systems. Herbicides and fungicides interfere with specific metabolic pathways in plants, such as photosynthesis and hormone function.

Pesticides pose a major threat of contamination to both surface water and ground water. In the United States, approximately 1 billion

---

pounds of pesticides are applied annually to crops.[34] Persistent and broad-spectrum agents such as DDT were once favored. DDT was shown to accumulate in the food chain and in living systems, with profound effects, and was prohibited in the United States in 1972; however, it and related chlorinated compounds continue to be used widely outside North America. Moreover, the nonresidual and more specifically targeted chemicals and agents that are now in wide use in North America still generate concern because of their long-term effects on ground and surface water.

Highly water-soluble pesticides and herbicides can leach into groundwater; the less soluble, more persistent chemicals can be carried in surface-water runoff to lakes and streams.[34] More than 70 pesticides have been detected in groundwater.[34] Specific chemicals, such as atrazine, are still routinely detected in aquifers and wells.[35]

The most recognized hazard of pesticide exposure is the development of acute toxic effects at high levels of exposure, such as might be sustained by an agricultural worker. The health effects of low-level or prolonged pesticide exposures via drinking water are much less clear. Extrapolation of results from *in vitro* studies to humans suggests the possibility of incrementally increased risk of cancer for many of the pesticides in use. Epidemiological correlations have been found between elevated serum DDT plus DDE, its major metabolite, and subjects who reported hypertension, arteriosclerosis, and diabetes in subsequent years.[36] Of particular concern are recent findings that demonstrate a strong association between breast cancer in women and elevated serum levels of DDE.[55] The overall database of human epidemiological data is sparse, however. In addition, in view of the slower elimination of pesticides in humans and their greater life span, extrapolating toxicity data from experiments on animals to humans may underestimate risks.

The case of aldicarb, a pesticide that has been used widely in recent times in the United States, is illustrative of contemporary issues related to pesticides and groundwater. A carbamate insecticide, aldicarb has been used on a number of crops, including potatoes, which are grown in sandy soil. The combination of the chemical's being applied to soil rather than to plant leaves and the permeability of sandy soil has led to widespread groundwater contamination.[37] Aldicarb has been detected in groundwater in Maine, Massachusetts, New York, Rhode Island, Wisconsin, and other states.

The most sensitive health effect of aldicarb exposure is related to cholinesterase depression.[38] Goldman et al. estimated that doses of 0.002–0.0086 mg/kg caused symptoms related to cholinesterase depression among people eating watermelons and cucumbers contaminated with aldicarb.[39]

Fiore et al. found that women drinking aldicarb-contaminated groundwater sustained cellular changes suggestive of an adverse effect on the immune system.[40] Mirkin et al. found similar results in a clinical epidemiological study.[41] Whether these alterations in immune function are biological markers predictive of other toxic outcomes, such as an increased risk of cancer or infection, is currently unknown.

Volatile Organic Compounds

Other very common groundwater contaminants include halogenated solvents[42,43] and petroleum products, collectively referred to as *volatile organic compounds* (VOCs). Both groups of chemical compounds are used in large quantities in a variety of industries. Among the most common uses of the halogenated solvents are as ingredients in degreasing compounds, dry-cleaning fluids, and paint thinners. Military dumps have recently been recognized for their widespread environmental contamination with solvents.

Historically, once used, these chemicals were discharged directly to land, given shallow burial in drums, pooled in lagoons, or stored in septic tanks. Sometimes the sites were landfills situated over relatively impermeable soils or impoundments lined with impenetrable material; often, however, the sites were in permeable soils, over shallow water tables, or near drinking-water wells. Petroleum products frequently were stored in underground tanks that would erode, or were spilled onto soil surfaces.

These compounds are major contaminants in recognized hazardous-waste sites. For instance, of the 20 chemicals most commonly detected at sites listed on the EPA's National Priority List, 11 were VOCs: trichloroethylene, toluene, benzene, chloroform, tetrachloroethylene, 1,1,1,-trichloroethane, ethylbenzene, trans-1,2-dichloroethane, xylene, dichloromethane, and vinyl chloride.[44]

Unfortunately, the chemical and physical properties of VOCs allow them to move rapidly into groundwater. Almost all of the above chemicals have been detected in groundwater near their original sites, some reaching maximum concentrations in the hundreds to thousands of parts per million.[45] Once in groundwater, their dispersion is dependent on a number of factors, such as aquifer permeability, local and regional groundwater flow patterns, chemical properties, and withdrawal rates from surrounding groundwater wells.[46]

At high levels of exposure, VOCs can cause headache, impaired cognition, hepatitis, and kidney failure; at the levels of exposure most commonly associated with water contamination, however, cancer and reproductive effects are of paramount concern. Many of these compounds have been found to cause cancer in laboratory animals.

In 1986, Lagakos et al. found a significant association between access to water contaminated by halogenated solvents and an increased incidence of childhood leukemia, perinatal death, two of five categories of congenital anomalities, and two of nine categories of childhood disorders.[47] Others have questioned whether these associations are meaningful.[48] The study by Lagakos and many similar subsequent studies are confronted by a number of methodological obstacles that are typical of epidemiological research on toxic issues in communities. Such studies, for instance, have difficulty in determining individual exposure patterns, subjects who can be properly used as controls, and accurate information on morbidity and mortality.

In the meantime, regulatory efforts have depended largely on extrapolations from chronic-exposure studies in laboratory animals, in which carcinogens are essentially regulated in terms of concentrations leading to an "acceptable" amount of risk. It is to be hoped that future regulatory efforts will be based on better information (in the form of epidemiological and toxicological studies) and on a better understanding of what constitutes an "acceptable" amount of risk.

Radioactive Contamination

Small quantities of radioactive substances are commonly found in drinking water. Of most concern is radon, which can be found in groundwater in association with geologic formations which are rich in this isotope. Ingestion of radon is probably of little consequence, since it is poorly absorbed and since the alpha radiation it emits is very weakly penetrating; however, the aerosolization of water containing radon, such as occurs in a shower, can lead to inhalation and increased risk of lung cancer. Standards have been set for the amount of radioactivity, including that from radon, that is allowable in drinking water.[49] Water supplies in many communities, including some in the United States, probably exceed this limit.

Of additional concern for some regions of the world are the contamination of surface water and groundwater by radioactive compounds generated by the production of nuclear weapons and by the processing of nuclear fuel. Government secrecy has prevented recognition of many of these incidents for years. For instance, it was only recently acknowledged that radioactive wastes, including plutonium, were dumped directly into the Techa River from 1947 through 1956 at the USSR's weapons-production facility in Chelyabinsk.[50] The Techa was the main source of potable water for thousands of residents of villages along its shores. Similar contamination of surface and ground water has occurred at weapons-production sites in the United States.

Hazardous-Waste Sites and Groundwater Contamination

Many of the specific hazards discussed above threaten water supplies because of their presence at hazardous-waste sites. Epidemiological studies of communities near hazardous-waste sites are plagued by a number of methodological obstacles, some of which were mentioned in the preceding section. Even if studies are performed flawlessly, and an association is discovered, causality is far from proven; moreover, the complex mixtures of chemicals found at most hazardous-waste sites make it exceedingly difficult to pinpoint the culprit substance(s).

Nevertheless, such studies are vitally important. They provide information on the scope of the problem, and they serve to educate communities about the hazards and the possible (if not exact) risks. Moreover, methods of exposure assessment and outcome ascertainment are constantly improving, as is demonstrated by a recent study in which slight but significant increases in malformation rates were associated with residential proximity to hazardous-waste sites in New York State.[51]

Issues Related to Water Treatment and Use

Remedial action for a contaminated aquifer is complicated, time-consuming, expensive, and often not feasible. If a contamination plume is shallow and in unconsolidated material, excavation and removal is a possible solution; other strategies include *in situ* detoxification, stabilization, immobilization, and barrier formation. Similarly, decontamination of a surface water supply is often complicated by the multiplicity of contaminants involved. Methods of water treatment that might be employed include reverse osmosis, ultrafiltration, use of ion-exchange resins, and filtration through activated charcoal. Clearly, the best solution to the contamination of groundwater or surface water is prevention.

Methods used for disinfecting drinking water can have toxic effects, due to the disinfectants or by their by-products. In the United States chlorine is routinely used, because of its powerful and long-lasting antimicrobial effect and its low cost; however, as a by-product of chlorination, chlorine reacts with substances commonly found in water to generate trihalomethanes (THM), such as chloroform, which increase the risk of cancer. As a volatile organic compound, chloroform can be significantly absorbed through skin contact and inhalation during a shower.[52] Treatment with chloramine or ozone instead of chlorine eliminates THM formation but is more expensive. Chlorination has also been recently implicated in the formation of nonvolatile polar furanone compounds that are powerfully mutagenic.[53]

Contamination, water treatment, and expense must be considered in the context of usage patterns. In developed countries, high-quality water is used in huge quantities. In the United States, 50 gallons of high-quality water are consumed per capita per day for domestic uses alone (165 gallons, if one counts commercial uses as well). Less than 1 gallon is actually consumed; the rest is utilized in a myriad of activities, most of which do not require high quality. Approaches to decreasing the use of high-quality water include increased attention to methods of conservation and the institution of dual water systems in which separate plumbing systems deliver high-quality water for culinary use and less pure water for other uses.

Summary

The number of different industrial and agricultural chemicals that threaten public and private water supplies is enormous. Nitrates, heavy metals, pesticides, and volatile organic compounds are of most concern in terms of human health. The exact nature of the health risks from many of these exposures is not known; this is particularly true with respect to the relationship of low-level chronic exposures to cancer and other long-term effects. Additional epidemiological and toxicological research is important, as are improving risk-assessment methods[54] and defining societal notions of "acceptable" risk. Of equal importance, however, is using existing research to target the prevention of additional contamination of this resource that is so critical to health and survival.

References

1. *Drinking Water and Health.* National Academy of Sciences, 1977.

2. National Water Quality Inventory: 1988 Report to Congress. Environmental Protection Agency report EPA 440-4-90-003.

3. Egboka, B. C., et al. Principles and problems of environmental pollution of groundwater resources with case examples from developing countries. *Environmental Health Perspectives* 83 (1989): 39–68.

4. *Standard Methods for the Examination of Water and Waste Water,* seventeenth edition. American Public Health Association, 1980.

5. *Control of Communicable Diseases in Man,* fifteen edition. American Public Health Association, 1990.

6. Bennett, J. V., and Brachman, P. S., eds. *Hospital Infections.* Little, Brown, 1992.

7. *Nitrates: An Environmental Assessment.* National Academy of Sciences, 1978.

8. *The Nation's Water Resources, 1975–2000,* volume 1. Water Resources Council, 1978.

9. Agency for Toxic Substances and Disease Registry. *Public Health Assessment Guidance Manual.* U.S. Department of Health and Human Services, 1992.

10. *The Health Effects of Nitrate, Nitrite, and N-Nitroso Compounds.* National Academy of Sciences, 1981.

11. Environmental Protection Agency, Office of Drinking Water Health Advisories. Nitrate and nitrite. *Reviews of Environmental Contamination and Toxicology* 107 (1988).

12. Johnson, C. J., and Kross, B. C. Continuing importance of nitrate contamination of groundwater and wells in rural areas. *American Journal of Industrial Medicine* 18 (1990): 449–456.

13. Craun, G. F., Greathouse, D. G., and Gunderson, D. H. Methemoglobin levels in young children consuming high nitrate well water in the United States. *International Journal of Epidemiology* 10 (1981), no. 4: 309–317.

14. Shearer, L. A., et al. Methemoglobin levels in infants in an area with high nitrate water supply. *American Journal of Public Health* 62 (1972), no. 9: 1174–1180.

15. Beresford, S. A. A. Is nitrate in the drinking water associated with the risk of cancer in the urban UK? *International Journal of Epidemiology* 14 (1985), no. 1: 57–63.

16. Forman, D. Are nitrates a significant risk factor in human cancer? *Cancer Survey* 8 (1989): 443–458.

17. Dorsch, M. M., et al. Congenital malformations and maternal drinking water supply in rural South Australia: A case control study. *American Journal of Epidemiology* 119 (1984), no. 4: 473–486.

18. Arbuckle, T. E., et al. Water nitrates and CNS birth defects: A population-based case-control study. *Archives of Environmental Health* 43 (1988), no. 2: 162–167.

19. Fan, A. M., Willhite, C. C., and Book, S. A. Evaluation of the nitrate drinking water standard with reference to infant methemoglobinemia and potential reproductive toxicity. *Regulatory Toxicology and Pharmacology* 7 (1987): 135–148.

20. Nriagu, J. O., and Pacyna, J. M. Quantitative assessment of worldwide contamination of air, water, and soils by trace metals. *Nature* 333 (1988): 134–139.

21. *ATSDR Biennial Report to Congress,* volume 2. Agency for Toxic Substances and Disease Registry, 1988.

22. Environmental Protection Agency. 40 CFR Parts 141 and 142. Maximum contaminant level goals and national primary drinking water regulations for lead and copper; final rule. *Federal Register,* part II, June 1991, 56 (110): 26460–26564.

23. Centers for Disease Control. *Preventing Lead Poisoning in Young Children.* U.S. Department of Health and Human Services, 1991.

24. *The Nature and Extent of Lead Poisoning in Children in the United States: A Report to Congress.* Agency for Toxic Substances and Disease Registry, 1988.

25. Feinglass, E. J. Arsenic intoxication from well water in the United States. *New England Journal of Medicine* 288 (1973): 828–830.

26. Borgono, J. M., and Grieber, R. Epidemiological study of arsenicism in the city of Antofasgasta. In *Trace Substances in Environmental Health,* ed. D. D. Hemphill. University of Missouri, 1972.

27. Tseng, W. P., et al. Prevalence of skin cancers in the endemic area of chronic arsenicism in Taiwan. *Journal of the National Cancer Institute* 40 (1968): 453–463.

28. Health Hazards of Inorganic Arsenic. Occupational Safety and Health Administration, 1979.

29. Chen, C. L., Kuo, T. L., and Wu, M. M. Arsenic and cancers. *Lancet* 1 (1988): 414–415.

30. Bates, M. N., Smith, A. H., and Hopenhayn-Rich, C. Arsenic ingestion and internal cancers: A review. *American Journal of Epidemiology* 135 (1992): 462–476.

31. Nogawa, K., Kobayashi, E., and Honda, R. A study of the relationship between cadmium concentrations in urine and renal effects of cadmium. *Environmental Health Perspectives* 28 (1979): 161–168.

32. Bako, G., et al. The geographical distribution of high cadmium concentrations in the environment and prostatic cancer in Alberta. *Canadian Journal of Public Health* 73 (1982): 92–94.

33. Plimmer, J. R. Pesticide loss to the atmosphere. *American Journal of Industrial Medicine* 18 (1990): 461–466.

34. Ritter, W. F. Pesticide contamination of ground water in the United States—A review. *Journal of Environmental Science and Health* 25 (1990), no. 1: 1–29.

35. Sullivan, J. B., Jr., et al. Health-related hazards of agriculture. In *Hazardous Materials Toxicology: Clinical Principles of Environmental Health,* ed. J. B. Sullivan, Jr., and G. R. Krieger. Williams & Wilkins, 1992.

36. Igbedioh, S. O. Effects of agricultural pesticides on humans, animals, and higher plants in developing countries. *Archives of Environmental Health* 46 (1991): 218–223.

37. Moye, H. A., and Miles, C. J. Aldicarb contamination of groundwater. *Reviews of Environmental Contamination and Toxicology* 105 (1988): 99–146.

38. *Drinking Water and Health.* National Academy of Sciences, 1983.

39. Goldman, L. R., Beller, M., and Jackson, R. J. Aldicarb food poisonings in California, 1985–1988: Toxicity estimates for humans. *Archives of Environmental Health* 45 (1990), no. 3: 141–147.

40. Fiore, M. C., et al. Chronic exposure to aldicarb-contaminated groundwater and human immune function. *Environmental Research* 41 (1986): 633–645.

41. Mirkin, I. R., et al. Changes in T-lymphocyte distribution associated with ingestion of aldicarb-contaminated drinking water: A follow-up study. *Environmental Research* 51 (1990): 35–50.

42. Toxicological Profile for Trichloroethylene—Draft. Agency for Toxic Substances and Disease Registry, 1992.

43. Toxicological Profile for Tetrachloroethylene—Draft. Agency for Toxic Substances and Disease Registry, 1991.

44. Grisham, J. W., ed. *Health Aspects of the Disposal of Waste Chemicals.* Pergamon, 1986.

45. Upton, A. C., Knerp, T., and Toniol, P. Public health aspects of toxic chemical disposal sites. *Annual Review of Public Health* 10 (1989): 1–25.

46. Lowe, J. A. Groundwater contamination by volatile organic solvents. In *Hazardous Materials Toxicology: Clinical Principles of Environmental Health,* ed. J. B. Sullivan, Jr., and G. R. Krieger. Williams & Wilkins, 1992.

47. Lagakos, S. W., Wessen, B. J., and Zelin, M. An analysis of contaminated well water and health effects in Woburn, Massachusetts. *Journal of the American Statistical Association* 81 (1986): 583–614.

48. Marshall, E. Woburn case may spark explosion of lawsuits. *Science* 243 (1986): 418–420.

49. Cothern, C. R. Estimating the health risk of radiation in drinking water. *Journal of the American Water Works Association* 79 (1987): 45–51.

50. Hu, H., Makhijani, A., and Yih, K. *Plutonium: Deadly Gold of the Nuclear Age.* International Physicians Press, 1992.

51. Geschwind, S. A., et al. Risk of congenital malformations associated with proximity to hazardous waste sites. *American Journal of Epidemiology* 135 (1992): 1197–1207.

52. Maxwell, N. I., Burmaster, D. E., and Ozonoff, D. Trihalomethanes and maximum contaminant levels: The significance of inhalation and dermal exposures to chloroform in household water. *Regulatory Toxicology and Pharmacology* 14 (1991): 297–312.

53. Holmbom, B. Mutagenic compounds in chlorinated pulp bleaching waters and drinking waters. *IARC Science Publications* 104 (1990): 333–340.

54. Winter, C. K. Dietary pesticide risk assessment. *Review of Environmental Contamination and Toxicology* 127 (1992): 23–67.

55. Wolff, M. S., et al. Blood levels of organochlorine residues and risk of breast cancer. *Journal of the National Cancer Institute* 85 (1993): 648–652.

56. Ruff, H. A., et al. Declining blood lead levels and cognitive changes in moderately lead-poisoned children. *Journal of the American Medical Association* 269 (1993): 1641–1646.

Food Contamination due to Environmental Pollution

Elizabeth L. Bowen, M.D., Ed.D.
Howard Hu, M.D., M.P.H., Sc.D.

Human life is sustained by an environment that provides adequate food derived from plants, minerals, and animals. Whereas the spoilage of food and its contamination by infectious agents and their toxins have long been of concern in the field of public health, contamination by environmental pollution has been less well recognized. We will concentrate on the latter, and we will not discuss hazards related to food processing (such as food irradiation, food additives, cooking, and preservation techniques) or to natural food toxins.

Environmental contamination of food can occur through multiple pathways on land, by air, and in fresh and salt water. The polluting agents of most significance include pesticides, radionuclides, halogenated cyclic compounds, and heavy metals.

A great deal of overlap exists between contamination of food and contamination of drinking water (see the preceding chapter) with respect to the toxins involved and the sources of pollution. In particular, aquatic animals serve as important contributors to the nutritional protein, lipid, and vitamin requirements of humans, and serve to cycle water-borne anthropogenic toxic chemicals back to human consumers in the form of food.[1]

In general, one the main differences between water contamination and food contamination is the tendency of plants and animals in the food chain to concentrate certain toxins, thereby increasing the exposure of unwary consumers. For instance, radioactive strontium concentrates in milk, and mercury (as the organic compound methyl mercury) concentrates in the tissues of fish.

The toxicity of contaminants in food can be compounded by malnutrition. For example, children who are deficient in iron, calcium, phosphorus, protein, or zinc absorb more lead than do well-nourished children with identical environmental lead exposures.[2] And malnutrition weakens the immune system, thus making an affected person more vulnerable to infectious pathogens and possibly to chemical agents.

Exposure to environmental food contamination may not be borne equally. In the United States, approximately three-fourths of the toxic waste disposal sites that failed to comply with the regulations of the U.S. Environmental Protection Agency were located in impoverished communities of people of color,[3] placing them at greater risk of food and water contamination. These are also the individuals who are at greatest risk for malnutrition and occupational exposures to pesticides, toxic metals, and other hazardous substances.

Responsibility for monitoring and control of contaminants in food is shared by a number of agencies. In the United States, the Food and Drug Administration (FDA) monitors dietary intake of selected contaminants; the Food Safety Inspection Service of the Department of Agriculture monitors residues in meat and poultry; and the Environmental Protection Agency's National Human Monitoring Program estimates total body exposure to toxic substances, including pesticides. Elsewhere, a growing number of countries are participating in the Global Environment Monitoring System, a program of food monitoring supported by the World Health Organization and the United Nations. By 1988, 35 countries participated, representing countries in every continent.[63]

Pesticides

Pesticides are used in agriculture in all parts of the world. While most cases of acute, high-exposure pesticide poisoning are related to occupational exposure to the applicators themselves (there are more than 200,000 deaths worldwide each year, mainly in this population, from acute pesticide poisoning[4]), significant exposure can occur through ingestion of treated food. At least 37 epidemics directly due to pesticide contamination of food have been reported.[5]

The term *pesticides* includes insecticides, herbicides, rodenticides, food preservatives, and plant growth regulators. We will concentrate on chemical insecticides. Chemical insecticides include synthetic organic insecticides and inorganic chemicals (mostly metals, such as arsenic). Other insecticides, such as those from biological sources—nicotine, pyrethrin, phermones, and insect-specific bacteria and viruses—will not be considered in this chapter. Synthetic organic insecticides can be further broken down into the chlorobenzene derivatives (e.g., dichlorodiphenyltrichloroethane (DDT)), cyclodienes (chlordane, aldrin, dieldrin), benzenehexachlorides (lindane), carbamates, and organophosphates (malathion).

While the mechanism of action differs among different classes of agents, most chemical pesticides are designed to be acutely toxic to their target organism. At high levels of exposure, they are also acutely toxic to humans, usually causing general symptoms of poisoning (nausea, vomiting, malaise, headache) as well as neurological symptoms (excitability, tremors, convulsions). Pesticide applicators are most at risk for high levels of exposure.

Pesticide contamination of food is mostly of concern because, while exposures are at lower levels, they involve much larger segments of the population (all consumers). In addition, many pesticides concentrate in the food chain and can accumulate in human tissue, where their slow metabolism and solubility in adipose (fat) tissue can lead to lifelong storage. Organochlorine pesticides have been found throughout the food chain, even in zooplankton and fish in the Arctic Ocean.[6] One recent study in Asia found these same pesticides at particularly high levels in preserved fruits, eggs, and fish.[7] Another study in Africa found the presence of chlorinated pesticides in over 80% of samples of eggs, poultry liver, and bovine liver and kidney; 7.5% of samples had levels higher than international tolerance levels.[8] In the United States, the commercial milk supply in Hawaii was contaminated by heptachlor epoxide during 1981 and 1982.[9] Isomers of dioxin have been found in crustaceans and finfish off the east coast, probably as the result of a combination of municipal and industrial combustion processes.[10]

Lindane has been detected in the blood and adipose tissue of the general public in a number of countries, probably because of food contamination. In autopsy surveys conducted in the 1970s, lindane and other benzenehexachlorides were present in more than 90% of human adipose tissue samples at a level of around 300 ppb.[11,12] Benzenehexachlorides were also found to be present in 82% of human breast milk samples at a mean concentration of 81 ppb with a range of 0–480 ppb.[13]

With respect to low-level exposure to humans, the toxic outcomes of greatest concern are cancer, immunotoxicity, and reproductive effects. This is an area of great concern, given the potential for population-wide exposures to a wide variety of pesticide residues, and great dispute, given the lack of toxicity and epidemiological data on most of these substances.

Most of the data on pesticides derive from in vitro assays that test the potential of a chemical to alter the genetic material of bacteria, and from studies on rodents. In the United States, these tests are now being coordinated by the National Toxicology Program. *In vitro* assays, however, cannot reliably be related to humans. Animal studies typically test high doses of a chemical. Statistical methods for extrapolating the risk of low-dose exposure from high-dose tests vary widely, give widely differing results, and are another subject of debate.

Nevertheless, the Delaney Clause of the Federal Food, Drug and Cosmetics Act has mandated both the EPA and the FDA to specially target food additives, including pesticides, that have been found to induce cancer when ingested by either animals or humans. This complicated legislation sets out specific criteria for the designation of contaminants in foods as potential carcinogens; once designated, potential carcinogens are then banned from foods sold in interstate commerce.[64] Lending support to the Delaney clause is a new study which has shown a strong association between breast cancer in women and elevated levels of serum DDE, the main metabolite of DDT.[65]

The current status of several major pesticides and the complicated interplay between evidence of carcinogenicity, scientific standards of proof, and regulatory standards can be seen in table 1.

The approach developed by regulatory agencies in the United States has not been without controversy. In the debate on the setting of

Table 1 Evidence for pesticide carcinogenicity. Adapted, with permission, from Council on Scientific Affairs. "Cancer Risk of Pesticides in Agricultural Workers," *Journal of the American Medical Association* 260 (1988), no. 7: 959–966.

Compound	Evidence Animal	Human	*In vitro*	IARC*	EPA†
Aldrin	Limited	Inadequate	Inadequate	3	C
Amitrole	Sufficient	Inadequate	Inadequate	2B	B2
	Sufficient	Inadequate	Inadequate	3	C
α-Naphthyithiourea	Inadequate	Inadequate	. . .	3	C
Aramite	Sufficient	. . .	. . .	. . .	. . .
Arsenicals	Inadequate	Sufficient	Limited	1	A
Benzal chloride	Limited	Inadequate	Limited	3	C
Benzotrichloride	Sufficient	Inadequate	Limited	2B	B2
Benzoyl chloride	Inadequate	Inadequate	Inadequate	3	C
Benzyl chloride	Limited	Inadequate	Sufficient	3	C
Captan	Limited	Insufficient	. . .	. . .	. . .
Carbon tetrachloride	Sufficient	Inadequate	Inadequate	2B	B2
Chlordane	Limited	Inadequate	Inadequate	3	C
Chlordimeform	No data	Insufficient	. . .	. . .	. . .
(metabolite)	Sufficient	. . .	. . .	. . .	. . .
Chlorobenzilate	Limited	Insufficient	. . .	. . .	. . .
Chlorophenols	. . .	Limited	. . .	2B	B2
Chlorothalonil	Limited	Insufficient	. . .	. . .	. . .
Diallate	Limited	Insufficient	. . .	. . .	. . .
1,2-Dibromochloro-propane	Sufficient	. . .	. . .	. . .	. . .
p-Dichlorobenzene	Sufficient	No data	. . .	2B	B2
2-Dichloroethane	Sufficient	. . .	. . .	. . .	. . .
2,4-Dichlorophe-noxyacetic acid esters	Inadequate	Inadequate	Inadequate	3	C
p,p′-Dichlorodiphenyl-trichloroethane	Sufficient	Inadequate	Inadequate	2B	B2
Dicofol (Kelthane)	Limited	Insufficient	. . .	. . .	. . .
Dieldrin	Limited	Inadequate	Inadequate	3	C
Ethylene dibromide	Sufficient	Inadequate	Sufficient	2B	B2

Table 1 (continued)

Compound	Evidence			IARC*	EPA†
	Animal	Human	*In vitro*		
Ethylene oxide	Limited	Inadequate	Sufficient	2B	B2
	Sufficient	Inadequate	. . .	. . .	. . .
Ethylene thiourea	Sufficient	Inadequate	Limited	2B	B2
Fluometuron	Inadequate	No evaluation	. . .	. . .	. . .
Formaldehyde	Sufficient	Inadequate	Sufficient	2B	B2
Heptachlor	Limited	Inadequate	Inadequate	3	C
Hexachlorobenzene	Sufficient	. . .	. . .	. . .	. . .
Kepone (chlordecone)	Sufficient	. . .	. . .	. . .	. . .
Lindane (γ-hexachlorocyclohexane)	Limited	Inadequate	Inadequate	3	C
Malathion	No evidence	No data	. . .	. . .	. . .
(4-chloro-2-methylphenoxy) acetic acid	Inadequate	Inadequate	. . .	3	C
Methyl parathion	No evidence	No evidence	. . .	3	C
Mirex	Sufficient	. . .	. . .	. . .	. . .
Nitrofen	Sufficient	No data	. . .	. . .	. . .
Parathion	Inadequate	Insufficient	. . .	. . .	. . .
Pentachlorophenol	Inadequate	Inadequate	Inadequate	3	C
Phenoxy acids	. . .	Limited	. . .	2B	B2
o-Phenylphenol	Limited	Insufficient	. . .	. . .	. . .
Piperonyl butoxide	No evidence	No evidence	. . .	. . .	. . .
Sulfallate	Sufficient	No data	. . .	. . .	. . .
2,3,7,8-Tetrachlorodibenzo-p-dioxin	Sufficient	Inadequate	Inadequate	2B	B2
Tetrachlorovinphos	Limited	Insufficient	. . .	. . .	. . .
Thiourea	Sufficient	. . .	. . .	. . .	. . .
Toxaphene	Sufficient	. . .	. . .	. . .	. . .
Trichlorfon	Inadequate	Insufficient	. . .	. . .	. . .

Table 1 (continued)

Compound	Evidence Animal	Human	In vitro	IARC[*]	EPA[†]
2,4,5-Trichlorophenol	Inadequate	Inadequate	No data	3	C
2,4,6-Trichlorophenol	Sufficient	Inadequate	No data	2B	B2
2,4,5-trichlorophe-noxyacetic acid	Inadequate	Inadequate	Inadequate	3	C
Vinyl chloride	Sufficient	Sufficient	Sufficient	1	A

[*]IARC indicates International Agency for Research on Cancer. Evidence is divided into the following categories: 1, evidence is sufficient to establish a causal relationship between the agent and human cancer; 2, agent, or process, is probably carcinogenic to humans; 2A, limited, almost sufficient evidence for carcinogenicity in humans; 2B, combination of sufficient evidence in animals and inadequate human data; and 3, cannot be classified according to carcinogenicity in humans.

[†]EPA indicates Environmental Protection Agency. Evidence is divided into the following groups: A, carcinogenic to humans (epidemiologic evidence supports a causal relationship); B, probably carcinogenic to humans (B1, epidemiologic evidence is limited or the weight of evidence from animal studies is sufficient or B2, evidence is sufficient from animal studies but epidemiologic studies provide inadequate evidence or no data); and C, possibly carcinogenic to humans (limited evidence from animal studies and no human data).

standards and allowable risks, some have argued that the risks posed by man-made pesticides are dwarfed by those posed by "natural" carcinogens, such as aflatoxin, ochratoxin A, and other mycotoxins.[14,15] Clearly, the relative impact of naturally occurring carcinogens needs better assessment, and attempts are underway in several countries to regulate the levels of these substances in food.[16] Nevertheless, the consensus among most public health authorities has been that it is critically important to continue to tighten the regulation of man-made pesticides, even while scientific research continues.[17]

But pesticides that are banned in the United States specifically because of their toxicity are still being exported. According to a 1987 United Nations report,[18] the U.S. EPA identified a significant risk for cancer for 18 of the 72 most commonly used pesticides. Of these 72 pesticides, 20 have been banned, withdrawn, or severely restricted in at least one country. Many are still produced and exported by U.S. companies, however.

Aggressive marketing by producers of pesticides is at least partially responsible for their indiscriminate use in the developing world. Dangers inherent in the use of pesticides are often minimized. Alternatives are not encouraged. Conditions in the developing world, including illiteracy, lack of personal protective equipment, and inadequate sanitation, increase the risk of both occupational and food-borne pesticide exposure to workers, their children, and entire communities.[14]

Only recently has there been a movement away from indiscriminate use of pesticides and toward a philosophy of "integrated pest management." Numerous studies have demonstrated clear health advantages to discontinuing current methods of intensive pesticide use in favor of "sustainable farming" (natural pest control, crop rotation, organic fertilizers, etc.).[19]

Radioactive Fallout

Decades of testing, operations, and accidents involving nuclear weapons production and nuclear energy power plants have resulted in contamination of plants and animals important to the food chain. Exposure to food is of concern because of the potential for internal irradiation and the permanent incorporation of radioactive material in the molecules of cells. Many radionuclides are involved. Of most significance with regard to levels of exposure, half-life, and toxicity are strontium 90, cesium 137, zirconium 95, carbon 14, plutonium 239, and iodine 131. Milk and other livestock products are particularly at risk.

Radionuclides accumulate in plants and animals according to their biological significance or their chemical similarity to biologically active elements in the same chemical groups.[20] Kinetic models of radioactivity transfer suggest that the dose from fallout is most sensitive to levels of deposition and direct transfer from contaminated feed and pastures to milk or meat.[21] The consumption of contaminated marine and freshwater fish is also of concern.[22]

Radioactive fallout that has already occurred, principally from atmospheric testing of nuclear weapons, is expected to cause at least 430,000 cancer deaths by the end of the 20th century. Because of the long half-lives of many fallout radionuclides, the total toll in premature human deaths has been estimated at more than 2 million.[23]

Polyhalogenated Aromatic Hydrocarbons

Much publicity has been generated on the subject of environmental contamination by polychlorinated biphenyls (PCBs). In the 1970s these synthetic compounds, once widely used in electrical transformers, lubricants, sealants, television sets, and other household products, were found to be potent animal carcinogens and to concentrate in the food chain (particularly in food fish). They were phased out and banned in the United States.

Some epidemiological studies of populations occupationally exposed to PCBs have reported excesses of liver and biliary tree cancers[24] and hematologic neoplasms[25]; however, other studies have been negative. A review of epidemiological studies of humans exposed to PCB-contaminated fish from the Great Lakes suggested a consistent relationship between PCB exposure and indicators of impaired neonatal and early infant health.[26]

Despite the banning of PCBs, and despite a few positive epidemiological studies, concern remains. The U.S. National Human Adipose Tissue Survey found that levels of PCBs above 3 ppm in the general population have been decreasing; however, the prevalence of any detectable level has been increasing.[27] Many bodies of water, such as the Great Lakes, remain heavily contaminated by PCBs.

Other polyhalogenated aromatic hydrocarbons of concern include the polybrominated biphenyls (PBBs) and the chlorinated benzenes (the most common chlorobenzene compound is hexachlorobenzene), both of which become concentrated in the adipose tissues of animals and humans. However, except in Michigan (where PBBs were accidentally mixed into livestock feed), widespread contamination by PBBs has not been a problem. Very little is known about the toxicity of PBBs, although anecdotal cases linking them to leukemia and other hematologic disorders have been reported.[28]

Heavy Metals

Lead

The contamination of food with lead is of major concern because of the high levels of exposure experienced around the world and because of recent studies linking neurobehavioral toxicity to relatively minute quantities of lead in human tissues.

Lead is a frequent contaminant of water, which may be used to irrigate crops or to process food in factories or homes. Lead in glazed ceramic ware or in crystal glassware can leach out, particularly if the food or drink contained is acidic. (Glazed ceramic ware has been a particularly high contributor to population-wide lead exposure in Mexico.[29]) The combustion of leaded gasoline (which went on for decades) has led to fallout of lead oxide in dust, which can contaminate the soil used to grow crops and the feed of livestock and which also can contaminate foods directly. Home gardens can be contaminated by peeling chips or by rain-washed runoff from lead paint, a common interior and exterior coating for dwellings built before 1955. Up to 50% of U.S. housing has lead paint on exposed surfaces.[30] Green leafy vegetables, in particular, concentrate lead. The U.S. Agency for Toxic Substances and Disease Registry (ATSDR) estimates that a million U.S. children are exposed to enough lead in food to cause lead poisoning.[37]

In 1980, half of the food and beverage cans produced in the United States were lead-soldered. Now few domestic products are, but food and drink cans manufactured outside the United States typically continue to contain lead solder, which can leach into food, and many such products are imported and consumed. Lead compounds are also used deliberately in substantial quantities in a variety of traditional "folk" medicines and cosmetics. Such toxic exposures have been reported among residents of China, India, the Middle East, and South America, and among immigrants from those regions.[19]

Lead exposure through food also occurs occupationally. The National Institute for Occupational Safety and Health (NIOSH) estimates that more than a million U.S. workers in more than 100 different occupations are exposed to lead on the job. At the workplace, airborne lead dust settles on hands, food, water, clothing, and other objects, and can be inhaled, ingested, or carried home.[31]

Ingestion of leaded paint chips and lead-contaminated dust is the major route of exposure for U.S. children. While everyone inhales or ingests small amounts of dust, smoke, or soil (on the order of 10 milligrams per day), young children may ingest as much as 200 mg per day,[32] including lead paint dust and lead-contaminated soil, particularly in cities or near highways. While the United States has removed lead from most gasoline, in many countries leaded gas and its airborne and soil contaminants are the dominant source of lead exposure, especially for children.

As figure 1 illustrates, lead interferes with many functions and structures in humans. Of most concern has been epidemiologic evidence demonstrating a link between lead exposure and adverse affects on the indices of intelligence and neurobehavioral development in children.[34,35] Even modest amounts seem to exert effects. Recently, the U.S. Centers for Disease Control (CDC) set the maximal level recommended in a child's blood at 10 µg per deciliter. This level was arrived

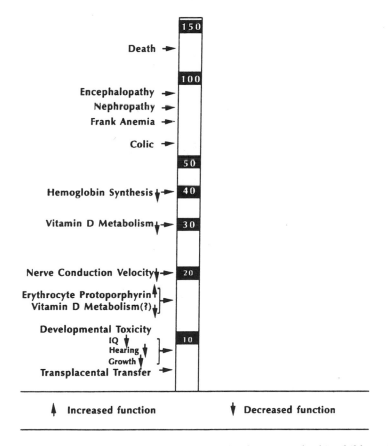

Figure 1 Lowest observed effect of levels of inorganic lead in children. (The levels are expressed in µg/dl of blood lead. They do not necessarily indicate the lowest levels at which lead exerts an effect; rather, these are the levels at which studies have adequately demonstrated an effect.) Source: reference 33.

at because of compelling evidence that there is no threshold to lead's harmful effect on intelligence. By this criterion, about one-sixth of the children in the United States have been estimated to have an excessive lead body burden that impairs health and interferes with the ability to learn.[36, 66] Fortunately, this cognitive and neurobehavioral damage may be partly reversible if blood lead levels are reduced over time.[67]

In addition, in the United States, an estimated 4 million women of child-bearing age are exposed to excessive environmental lead. Maternal and umbilical-cord blood lead levels of 10 µg/dl are associated with low birth weight and prematurity, conditions that expose an infant to a host of health and developmental risks. Major societal investments of resources in the near future will be necessary to protect those women's fetuses from *in utero* lead poisoning. They may be expected to bear some 10 million children at risk of lead poisoning at birth.[37]

Problems of similar or even greater magnitude confront many other nations, including those in Eastern Europe and the former Soviet Union. Responsible public-health decision makers are continually revising "safe" standards as more refined data become available. The same process will no doubt be true of many other toxins in our food, water, air, and homes about which we have far less knowledge than we do about lead.

Other Heavy Metals

Lead is not the only metal to contaminate food. Several highly toxic metals that are often used in agricultural and industrial applications may enter the food supply intentionally or inadvertently. Common domestic sources of exposure to arsenic, cadmium, copper, and mercury are pottery, metal pans, teapots, cooking utensils, and packaging materials. Arsenic, copper, and mercury are also used in herbicides, fungicides, and insecticides. Any and all of these routes of exposure can cause food contamination and can produce acute or chronic illnesses.

Cadmium

Cadmium can contaminate food by its presence in pesticides, pigments, paints, plastics, and cigarettes. In the United States, 500,000 individuals have occupational exposures to cadmium in mining, welding, galvanizing, battery production, and many other industries. Families who live near the sites of such industries or who are engaged in cottage industries involving cadmium-containing pigments or batteries may also develop cadmium toxicity through exposure to cadmium in food, air, soil, and water.[38] Substantial cadmium pollution can occur in areas where arsenic, zinc, copper, lead, and cadmium are mined from iron ore. In Japan, cadmium runoff from mines has polluted rivers that were used to irrigate rice paddies. Individuals who consumed cadmium-contaminated rice developed chronic cadmium poisoning and had shortened life spans.[39]

Cadmium accumulates throughout life. High exposure has been linked to osteomalacia, a softening of the bones.[40] Cadmium damages renal tubules, causing proteinuria, a condition in which serum proteins are excreted in excess in the urine. A dose-response relationship has been shown between the prevalence of proteinuria and the cadmium content of rice in contaminated regions.[41] Finally, substantial concern exists over the possibility, suggested by animal research and epidemiological studies, that chronic lower-dose cadmium exposure can cause cancer, particular of the lung and of the prostate.

Mercury

Mercury contamination of food has been well documented in locations as diverse as Michigan, Iraq, and Japan. A classic episode occurred in the 1950s in Minamata Bay, Japan. A chemical factory that made vinyl chloride dumped mercury into the bay. Individuals who ate contaminated fish developed mercury toxicity accompanied by neurological disorders, including progressive peripheral paresthesias with numbness and tingling sensations in the extremities, loss of muscle coordination with unsteadiness of gait and limbs, slurred speech, irritability, memory loss, insomnia, and depression. Forty deaths and at least 30 cases of cerebral palsy with permanent disability were reported.[42]

A much larger epidemic of similar neurological disorders occurred in Iraq when seed grain treated with mercury fungicide, instead of being planted, was mistakenly incorporated into wheat flour and baked into bread. More than 450 persons died, and more than 6,000 were hospitalized.[43]

In the United States, an estimated 68,000 workers are exposed to mercury in the workplace. The major agricultural and industrial sources of mercury are fungicides, pesticides, paints, pharmaceuticals, batteries, electrical equipment, thermometers, and the industrial production of chlorine and vinyl chloride.[44]

Ingestion of contaminated fish and fish products is a major source of environmental exposure to mercury. In the United States, mercury contamination of freshwater fish is prevalent in the Great Lakes region. Excessive levels of methylmercury have been reported in fish in scores of Michigan lakes. Public health authorities in 20 states have issued advisories that children, women of child-bearing age, and pregnant and lactating women should avoid eating certain fishes from contaminated lakes. However, an estimated 20% of the fish and shellfish consumed in the United States comes from subsistence fishing or recreational fishing and is not subject to adequate monitoring from an environmental health standpoint.[45]

Mercury compounds from agricultural and industrial sources are converted by bacteria into methylmercury, which is soluble, mobile, and rapidly incorporated into aquatic food chains. Mercury concentrates as it moves up the food chain, accumulating in carnivorous fishes (such as the northern pike) to levels 10,000–100,000 times the concentrations in the surrounding water.[46] Marine fishes, especially carnivorous ones such as the swordfish, have been found to contain high levels of mercury, exceeding 1 μg per gram.[47] Between 70% and 90% of the mercury detected in fish muscle is in the bioavailable form of methylmercury and hence is readily absorbed.[48]

Environmental agencies in New York, Wisconsin, and Minnesota have reported an association between lake acidification from acid rain and increasing levels of mercury in fish. Tropospheric ozone pollution and global warming may also lead to increased levels of mercury in freshwater fish, the former by increasing the rate of conversion of elemental mercury to methyl mercury[49] and the latter through increased atmospheric mercury deposition.[50] In-depth reviews of the subject of food-related mercury toxicity and safety can be found in references 51 and 52.

Arsenic

Arsenic is used widely in insecticides, fungicides, and herbicides, and may contaminate food by all these routes. Diet represents the largest source of arsenic exposure for the general population, followed by groundwater contamination. In addition, an estimated 55,000 U.S. workers have had occupational exposures to arsenic.

Arsenic is found in 28% of U.S. "Superfund" hazardous-waste sites, and migration from those sites, with subsequent contamination of food and water, has been documented.[53] And young children living near pesticide factories or copper smelters may ingest arsenic-contaminated soil on playgrounds, adding to the possibility of their developing arsenic toxicity.

Symptoms of acute arsenic toxicity are nausea, vomiting, diarrhea, abdominal pain, and metallic taste. Severe toxicity may cause circulatory collapse, seizures, and kidney failure due to acute tubular necrosis. Chronic exposure to moderately high levels of arsenic is associated with fatigue, weakness, gastroenteritis, dermatitis, and peripheral neuropathies that begin with painful feet and progress to a loss of normal sensation in the hands and feet in a "stocking and glove" pattern.

In Taiwan, chronic exposure to moderately elevated levels of arsenic in food or drinking water have been linked to an increased risk of skin cancer (and also, perhaps, to "black foot disease").[54] A positive dose-response relationship was also observed for bladder, liver, and lung cancers.[55]

The potential for carcinogenicity remains a primary concern for exposure to arsenic at low levels. As with many topics related to food toxicology, little epidemiological research exists which can address this issue; extrapolation from high-exposure studies using conventional methods suggests that significant risks may exist.

Copper

Copper is used widely in many industries, including agriculture; it is used in plumbing and in cookware; and has been identified in 18% of U.S. hazardous-waste sites. Acidic drinking water mobilizes copper from plumbing. In many countries, including the United States, copper sulfate is added directly to reservoirs to control algae. This sharply raises the level of copper in drinking water for several days.[52]

With very high levels of exposure, acute copper poisoning results in nausea, vomiting, diarrhea, and metallic taste. Chronic copper toxicity has been studied in the context of Wilson's disease, a rare inherited metabolic disease in which copper accumulation leading to central-nervous-system degeneration, liver disease, and anemia.[56] *In vitro* studies and mammalian *in vivo* studies suggest that copper may also be a human mutagen.[57] Relatively little is known about the potential toxicity of copper at the levels of exposure most commonly encountered. There is reason for concern, however, because of the very broad human exposure to copper compounds.

Miscellaneous Contamination

Food can be inadvertently contaminated by industrial chemicals mistakenly introduced during processing and distribution. For instance, the ingestion of refined aniline-adulterated rapeseed oil in Spain in 1981 was associated with the development of a toxic syndrome with autoimmunological features.[58]

The recent introduction of food irradiation has generated some concerns regarding the potential induction of harmful radioactivity, radiolytic products (such as superoxide radicals), and mutant strains of microorganisms.[59] Little hard evidence exists that supports these concerns.[60] Nevertheless, additional research seems prudent in view of the widespread potential application of this method of preserving food.

Conclusion

The integrity of food is threatened by a number of man-made pollutants that can be introduced at any step in the food chain and in the food-processing industry. There have been a number of instances in which high-level poisoning has occurred through human error and negligence. The potential toxicity of exposures to pesticides, metals, radionuclides, and other contaminants that have slowly accumulated in soils and the food chain is of growing concern.

Most of these toxins are invisible and are not easily detected by consumers. Moreover, the processing or cooking of food is generally not effective in neutralizing their impact. For instance, broiling fish contaminated with polychlorinated biphenyls and pesticides has not been found to significantly alter their levels.[61]

Painfully slow research has begun to clarify the risks associated with food contamination. New tools are being developed to better define accumulated exposure and early health effects in humans[62]; in the meantime, it would seem prudent to pursue primary prevention and to vigilantly guard against the contamination of the food supply by environmental pollutants.

References

1. Dawe, C. J. Implications of aquatic animal health for human health. *Environmental Health Perspectives* 86 (1990): 245–255.

2. Mahaffey, K. R., and Michaelson, I. A. The interaction between lead and nutrition. In *Low Level Lead Exposure: The Clinical Implications of Current Research*, ed. H. L. Needleman. Raven, 1980.

3. Kennedy, R. F., Jr., and Rivera, D. Pollution's chief victims: The poor. *New York Times*, August 15, 1992.

4. *Public Health Impact of Pesticides Used in Agriculture*. World Health Organization, 1990.

5. Ferrer, A., and Cabral, J. P. R. Epidemics due to pesticide contamination of food. *Food Additives and Contaminants* 6 (1989), suppl. 1: S95–S98.

6. Hargrave, B. T., et al. Organochlorine pesticides and polychlorinated biphenyls in the Arctic Ocean food web. *Archives of Environmental Contamination and Toxicology* 22 (1992): 41–54.

7. Ip, H. M. Chlorinated pesticides in foodstuffs in Hong Kong. *Archives of Environmental Contamination and Toxicology* 19 (1990): 291–296.

8. Kessabi, M., et al. Contamination of eggs, poultry liver, and bovine liver and kidney by chlorinated pesticides in Morocco. *Science of the Total Environment* 90 (1990): 283–287.

9. Baker, D. B., Loo, S., and Barker, J. Evaluation of human exposure to the heptachlor epoxide contamination of milk in Hawaii. *Hawaii Medical Journal* 50 (1991): 108–112.

10. Wenning, R. J., et al. Potential sources of 1,2,8,9-tetrachlorodibenzo-P-dioxin in the aquatic environment. *Ecotoxicology and Environmental Safety* 23 (1992): 133–146.

11. *Toxicological Profile for Hexachlorocyclohexane*. Agency for Toxic Substances and Disease Registry, 1989.

12. de Vlieger, M. et al. The organochlorine insecticide content of human tissues. *Archives of Environmental Health* 17 (1968): 759–767.

13. Van Ert, M., and Sullivan, J. B., Jr. Organochlorine pesticides. In *Hazardous Materials Toxicology—Clinical Principles of Environmental Health,* ed. J. B. Sullivan, Jr., and G. R. Krieger. Williams & Wilkins, 1992.

14. Ames, B. N., Profet, M., and Gold, L. S. Dietary pesticides (99.989% all natural). *Proceedings of the National Academy of Sciences* 87 (1990): 7777–7781.

15. Scheuplein, R. J. Perspectives on toxicological risk—an example: Foodborne carcinogenic risk. *Critical Review of Food Science and Nutrition* 32 (1992): 105–121.

16. Stoloff, L., Van Egmond, H. P., and Park, D. L. Rationales and the establishment of limits and regulations for mycotoxins. *Food Additives and Contamination* 8 (1991): 213–221.

17. Fan, A. M., and Jackson, R. J. Pesticides and food safety. *Regulatory Toxicology and Pharmacology* 9 (1989): 158–174.

18. Consolidated List of Products Whose Consumption and/or Sale Have Been Banned, Withdrawn, Severely Restricted or Not Approved by Governments. United Nations publication ST/ESA/192, 1987.

19. Vogtmann, H. From healthy soil to healthy food: An analysis of the quality of food produced under contrasting agricultural systems. *Nutrition and Health* 6 (1988), no. 1: 21–35.

20. Poston, T. M., and Klopfer, D. C. Concentration factors used in the assessment of radiation dose to consumers of fish: A review of 27 radionuclides. *Health and Physiology* 55 (1988): 751–766.

21. Prohl, G., Muller, H., and Voigt, G. The influence of the feeding practice and the season on the contamination of animal food products after a single deposition of radionuclides. *Science of the Total Environment* 85 (1989): 107–117.

22. Poston, T. M., and Klopfer, D. C. Concentration factors used in the assessment of radiation dose to consumers of fish: A review of 27 radionuclides. *Health and Physiology* 55 (1988): 751–766.

23. Robbins, A., Makhijani, A., and Yih, K. *Radioactive Heaven and Earth*. Apex, 1991.

24. Brown, D. P. Mortality of workers exposed to polychlorinated biphenyls—an update. *Archives of Environmental Health* 42 (1987): 333–339.

25. Bertazzi, P. A., et al. Cancer mortality of capacitor manufacturing workers. *American Journal of Industrial Medicine* 11 (1987): 125–176.

26. Swain, W. R. Effects of organochlorine chemicals on the reproductive outcome of humans who consumed contaminated Great Lakes fish: An epidemiologic consideration. *Journal of Toxicology and Environmental Health* 33 (1991): 587–639.

27. Mack, G. A., and Mohadjer, L. Baseline estimates and time trends for beta-benzene hexachloride, hexachlorobenzene, and polychlorinated biphenyls in human adipose tissue 1970–1983. U.S. Environmental Protection Agency, 1985.

28. Zapata-Gayón, C., Zapata-Gayón, N., and González-Angulo, A. Clastogenic chromosomal aberrations in 26 individuals accidentally exposed to orthodichlorobenzene vapors in the national medical center in Mexico City. *Archives of Environmental Health* 37 (1982): 231–235.

29. Hernández-Avila, M. et al. Lead-glazed ceramics as major determinants of blood lead levels in Mexican women. *Environmental Health Perspectives* 94 (1991): 117–120.

30. Chisholm, J. J. Removal of lead paint from old housing: The need for a new approach. *American Journal of Public Health* 76 (1986), no. 3 236–237.

31. Criteria for a Recommended Standard. Occupational Exposure to Inorganic Lead, Revised Criteria. Publication 78-158, National Institute for Occupational Safety and Health 1978

32. Calabrese, E. J., et al. Preliminary adult soil ingestion estimates: Results of a pilot study. *Regulatory Toxicology and Pharmacology* 12 (1990): 88–95.

33. Case Studies in Environmental Medicine: Lead Toxicity. Agency for Toxic Substances and Disease Registry, 1990.

34. Air Quality Criteria for Lead. Report EPA-600/8-83/028aF-dF, Environmental Protection Agency, 1986.

35. Grant, L. D., and Davis, J. M. Effects of low-level lead exposure on pediatric neurobehavioral development: Current findings and future directions. In *Lead Exposure and Child Development,* ed. M. A. Smith, L. D. Grant, and A. I. Sors. Kluwer, 1989.

36. Preventing Lead Poisoning in Young Children. Centers for Disease Control, 1991.

37. *The Nature and Extent of Lead Poisoning in Children in the United States.* Agency for Toxic Substances and Disease Registry, 1988.

38. Cadmium toxicity. In *Case Studies in Environmental Medicine.* Agency for Toxic Substances and Disease Registry, 1990.

39. Nakagawa, H., et al. High mortality and shortened life-span in patients with Itai-Itai Disease and subjects with suspected disease. *Archives of Environmental Health* 45 (1990): 283–287.

40. Friberg, L., et al., eds. *Cadmium and Health: A Toxicological and Epidemiological Appraisal, II. Effects and Response.* CRC Press, 1985.

41. Nogawa, K., and Ishizaki, A. A comparison between cadmium in rice and renal effects among inhabitants of the Jinzu river basin. *Environmental Research* 18 (1979): 410–420.

42. Agocs, M. M., et al. Mercury exposure from interior latex paint. *New England Journal of Medicine* 323 (1990): 1096–1100.

43. Bakir, F., et al. Methylmercury poisoning in Iraq. *Science* 181 (1973): 230–240.

44. *National Occupational Exposure Survey (1980–1983) Database.* Department of Health and Human Services, 1984.

45. Institute of Medicine. *Seafood Safety.* National Academy Press, 1991.

46. Environmental Protection Agency. Identification and listing of hazardous waste, discarded commercial products, off-specification species, container residues, and spill residues thereof. *Federal Register* 45 (1980).

47. Mercury Health Effects Update; Health Issues Assessment; Final Report. Report 600 8-84-019F, Environmental Protection Agency, 1984.

48. *Environmental Health Criteria: Mercury.* World Health Organization, 1976.

49. Schneider, K. Ancient hazards of mercury re-emerge. *New York Times,* August 26, 1991.

50. Fitzgerald, W. F., and Clarkson, T. W. Mercury and monomethylmercury: Present and future concerns. *Environmental Health Perspectives* 96 (1991): 159–166.

51. Institute of Medicine. *Seafood Safety.* National Academy Press, 1991.

52. National Research Council. *Animals as Sentinels of Environmental Health Hazards.* National Academy Press, 1991.

53. *ATSDR Biennial Report to Congress: Volume II.* Agency for Toxic Substances and Disease Registry, 1988.

54. Tseng, W. P. Effects and dose-response relationship of skin cancer and Blackfoot disease with arsenic. *Environmental Health Perspectives* 19 (1977): 109–199.

55. Chen, C. J., et al. A retrospective study of malignant neoplasms of bladder, lung, and liver in a Blackfoot disease endemic area in Taiwan. *British Journal of Cancer* 53 (1986): 399–405.

56. Medalia, A., and Scheinberg, I. H. Psychopathology in patients with Wilson's Disease. *American Journal of Psychology* 146 (1989), no. 5: 662–664.

57. Bhunya, S. P., and Pati, P. C. Genotoxicity of an inorganic pesticide, copper sulphate, in mouse *in vivo* test system. *Cytologia* 52 (1987): 801–808.

58. Kammuller, M. E., Bloksma, N., and Seinen, W. Chemical-induced autoimmune reactions and Spanish toxic oil syndrome. Focus on hydantoins and related compounds. *Journal of Toxicology and Clinical Toxicology* 26 (1988): 157–174.

59. McGovney, W. T. Preservation of food products by irradiation. *Seminar on Nuclear Medicine* 18 (1988): 36–45.

60. Swallow, A. J. Wholesomeness and safety of irradiated foods. *Advances in Experimental Medical Biology* 289 (1991): 11–31.

61. Trotter, W. J., et al. Levels of polychlorinated biphenyls and pesticides in bluefish before and after cooking. *Journal of the Association of the Official Annals of Chemistry* 72 (1989): 501–503.

62. Ozonoff, D., and Longnecker, M. P. Epidemiological approaches to assessing human cancer risk from consuming aquatic food resources from chemically contaminated water. *Environmental Health Perspectives* 90 (1991): 141–146.

63. *Global Environment Monitoring System: Assessment of Chemical Contaminants in Food. Report on the Results of the UNEP/FAO/WHO Programme on Health-Related Environmental Monitoring,* United Nations Environment Programme, Food and Agriculture Organization, World Health Organization, 1988.

64. Committee on Scientific and Regulatory Issues Underlying Pesticide Use Patterns and Agricultural Innovation, National Research Council. *Regulating Pesticides in Food: The Delaney Paradox.* National Academy Press, 1987.

65. Wolff, M. S., et al. Blood levels of organochlorine residues and risk of breast cancer. *Journal of the National Cancer Institute* 85 (1993): 648–652.

66. Needleman, H. L., et al. The long-term effects of exposure to low doses of lead in children. *New England Journal of Medicine* 322 (1990): 83–88.

67. Ruff, H. A., et al. Declining blood lead levels and cognitive changes in moderately lead-poisoned children. *Journal of the American Medical Association* 269 (1993): 1641–1646.

Occupational Exposures and Human Health

Dean B. Baker, M.D., M.P.H.
Philip J. Landrigan, M.D.

adapted, with permission, from an article by these authors in *Medical Clinics of North America* (74 (1990), no. 2: 441–460)

Workers may be exposed to high concentrations of toxic agents, and they may be exposed to these agents earlier than members of the general population. Also, workers typically constitute well-defined groups; the nature and extent of their exposures are known; and they are susceptible to epidemiologic analysis. Therefore, many environmentally induced diseases have first been seen in working populations. The appearance of these illnesses may provide a warning to the general population of the toxicity of environmental toxins. At the same time, occupational disease is not limited to the workplace. Toxic hazards may be released from the workplace into the community environment to pollute the air, the drinking water, or the food chain.[1] Also, occupational toxins may be transported home on the clothing of contaminated workers to cause such illnesses as lead poisoning and mesothelioma in family members.[2,3]

Occupational diseases encompass a broad range of human illness, including lung cancer and mesothelioma in asbestos workers, cancer of the bladder in dye workers, leukemia in workers exposed to benzene, chronic bronchitis in workers exposed to dusts, disorders of the nervous system in workers using solvents, kidney failure in lead workers, impairment of reproductive function in men and women using lead and some pesticides, and chronic disease of the musculoskeletal system in workers who suffer repetitive trauma.

Occupational diseases are underdiagnosed, and many diseases of occupational origin are incorrectly attributed to other causes. The underdiagnosis of occupational illness reflects the fact that, clinically and pathologically, most occupational illnesses are not distinct from chronic diseases associated with non-occupational etiologies. For example, lung cancer caused by asbestos is identical in its clinical and pathologic presentation to lung cancer caused by cigarette smoking.[4] Similarly, solvent-induced encephalopathy may be attributed to old age.[5] Only in rare instances, such as in the association between asbestos and mesothelioma[4] or in that between vinyl chloride monomer and angiosarcoma of the liver,[6] is the causal association between occupational exposure and disease established on clinical grounds alone.

The typically long latency period between occupational exposure and appearance of illness is another barrier to diagnosis. Added to this is the fact that many workers have had multiple toxic exposures and, at least until recently, have not been informed of the nature or hazard of the materials with which they worked.

Finally, the underdiagnosis of occupational disease reflects the fact that most physicians are not adequately trained to suspect work as a cause of disease. Very little time is devoted in most medical schools to teaching physicians to take a proper occupational history, to recognize the symptoms of common industrial toxins, or to recall known associations between occupational exposure and disease.[7] The average medical student receives less than 4 hours of training in occupational medicine during the 4 years of medical school.[8] Largely in consequence of this lack of training, most physicians do not routinely obtain histories of occupational exposure from their patients. Recent surveys indicate that adequate occupational histories are recorded on fewer than 10% of medical charts.[7]

The extent of underdiagnosis of occupational illness is illustrated by the fact that in 1982 only three cases of occupationally induced cancer were diagnosed as work-related and received compensation in New York State through the State Workers' Compensation Board. By contrast, estimates indicate that 2,000 to 3,000 deaths from occupationally induced cancer occur in New York State each year.[9] At least 1,000 of these deaths are the consequence of past occupational exposure to asbestos, a well-known carcinogen about which every physician should routinely inquire.[10]

A major challenge to the medical practitioner is, therefore, to properly diagnose occupationally induced disease.[7,11] Not only will proper diagnosis of occupational illness assist in the treatment and (if necessary) the compensation of the individual patient; it also should serve as a trigger for public health intervention to prevent disease in other, similarly exposed workers.[12]

Extent of Occupational Disease

The 1972 President's Report on Occupational Safety and Health estimated that 390,000 new cases of work-related disease occur annually and that 100,000 deaths result per year from occupational illnesses.[13] Although these numbers are imprecise, they may be taken as indicating the order of magnitude of the problem.

To develop a more reliable estimate of the burden of occupational disease in New York State, data were compiled from the state's workers' compensation program, from employers' reports to the U.S. Occupational Safety and Health Administration (OSHA), from reports to disease registries maintained by the State Department of Health, and from physicians' reports.[9] From these data it was estimated that between 4,700 and 6,600 deaths occur each year in New York State as the results of occupational exposure. In this analysis, 10% of cancer deaths, 100% of deaths from the pneumoconioses, and 1–3% of deaths from cardiovascular, chronic respiratory, renal, and neurologic disease were attributed to occupational exposure.[9] This range of estimates is conservative, reflecting the fact that proportional risks were applied to deaths in only six categories. Further, in each instance, conservative attributable risk fractions were selected from the range of estimates available in the literature.

A national estimate of the burden of occupational disease may be developed by multiplying the New York State data by 10; New York contains slightly less than 10% of the nation's workforce, and it includes a diverse mix of employment in the manufacturing, service, and agricultural sectors, thus representing a microcosm of employment in the United States. On this basis, it may be estimated that occupational disease is responsible each year for 50,000–70,000 deaths in the United States, and for approximately 350,000 new cases of illness. These numbers are not far from the national estimates developed more than a decade earlier by the President's Commission.[13]

Principal Types of Occupational Disease

Occupational illness can affect virtually every organ system. It is beyond the scope of this chapter to review the full range of occupational illnesses; several textbooks on occupational medicine are available.[14–16] To develop a systematic approach to the prevention of illness of occupational origin and to rank categories of occupational disease according to relative priority, the National Institute for Occupational Safety and Health (NIOSH) has developed a list (see table 1) of the ten most important categories of occupational illness.[17,18]

The examples on the next few pages will illustrate some major occupational diseases.

Table 1   NIOSH list of ten leading work-related diseases and injuries.* Source: reference 18.

1.   Occupational lung diseases:
     asbestosis, byssinosis, silicosis, coal workers' pneumoconiosis, lung cancer, occupational asthma

2.   Musculoskeletal injuries:
     disorders of the back, trunk, upper extremity, neck, lower extremity; traumatically induced Raynaud's phenomenon

3.   Occupational cancers (other than lung):
     leukemia; mesothelioma; cancers of the bladder, nose, and liver

4.   Severe occupational traumatic injuries:
     amputation, fracture, eye loss, laceration, and traumatic death

5.   Occupational cardiovascular diseases:
     hypertension, coronary artery disease, acute myocardial infarction

6.   Disorders of reproduction:
     infertility, spontaneous abortion, teratogenesis

7.   Neurotoxic disorders:
     peripheral neuropathy, toxic encephalitis, psychoses, extreme personality changes (exposure-related)

8.   Noise-induced loss of hearing

9.   Dermatologic conditions:
     dermatoses, burns (scaldings), chemical burns, contusions (abrasions)

10. Psychologic disorders:
     neuroses, personality disorders, alcoholism, drug dependency

*The conditions listed under each category are to be viewed as selected examples, not comprehensive definitions of the category.

Occupational Lung Diseases

Occupational lung diseases are extremely important in the field of occupational medicine, since the lung is both a route of entry and a target organ for toxic substances. Major categories of lung diseases include lung cancer, pneumoconioses, occupational asthma, industrial bronchitis, and infections. The number of potentially exposed workers is large. NIOSH estimates that there are over 1.2 million workers potentially exposed to silica dust alone.[19] Nicholson et al.[10] estimated that, from 1940 through 1979, some 27.5 million individuals in the United States had potential exposure to asbestos at work. Since the latency for asbestos-related lung cancer and mesothelioma typically is 25 years or more, the future burden of mortality due to asbestos would be substantial even if all future exposure were to be completely eliminated.

Occupational Cancer

Estimates of the fraction of cancers due to occupational exposure vary from 4% to 38%. Even the lowest of these estimates suggests that 17,000 cancer deaths per year in the United States are attributable to workplace exposure.[17] NIOSH's National Occupational Exposure Survey estimated that between 3 million and 9 million workers currently are potentially exposed to chemicals considered by NIOSH to be proven or likely carcinogens. No overall estimates exist of the size of the population with past exposure to occupational carcinogens.

Occupational Skin Disorders

Occupational skin disorders are important causes of morbidity and disability in the workplace. In 1987, the estimated rate of occupational skin disorders in the United States was 7.4 per 10,000 full-time workers.[18] Skin diseases account for 34–50% of all reported occupational diseases in the United States. However, in comparison with most other occupational health problems, skin disorders are relatively easily diagnosed and recognized as work-related. Occupational skin disorders are unevenly distributed among industries. A worker in agriculture, forestry, fishing, or manufacturing has 3 times as much risk of developing a work-related skin disease as workers in other industries.

Occupational Infectious Diseases

According to an estimate by the U.S. Centers for Disease Control, 12,000 health-care workers whose jobs entail exposure to blood become infected each year with Hepatitis B virus (HBV), over 500 of them are hospitalized as a result of that infection, and approximately 250 die (from fulminant hepatitis, cirrhosis, or liver cancer).[18] The potential for HBV transmission in the workplace is greater than that for human immunodeficiency virus (HIV), but the modes of transmission for these two viruses are similar. An increased risk of HIV infection must, therefore, be assumed to exist in occupational settings in which workers may be routinely exposed to blood or other bodily fluids.

Occupational Reproductive Disorders

There are relatively few firmly documented associations between occupational exposures and adverse reproductive outcomes, mostly because there have been few epidemiologic or clinical studies of sufficient quality.[20] Most knowledge about reproductive toxicity comes from laboratory studies. A recent review found that of 2,800 chemicals that have been evaluated for teratogenicity in animals, about 38% exhibit some teratogenic potential.[21] Current research is inadequate to estimate the contribution of occupational exposures to other reproductive disorders, such as reduced fertility, spontaneous abortion, and infant death.

Severe Occupational Traumatic Injuries

These include such events as amputation, fracture, severe laceration, eye loss, acute poisoning, and burns. NIOSH estimates that at least 10 million persons suffer traumatic injuries on the job each year. About 3 million of these injuries are severe. In 1985, the National Traumatic Occupational Fatality database—through which NIOSH collects death-certificate information on occupational-injury fatalities—recorded 6,442 occupational fatalities in the United States. The major causes of work-related deaths were highway motor-vehicle incidents (34%), falls (13%), non–highway industrial-vehicle incidents (11%), blows other than by vehicles or equipment (8%), and electrocutions (7%). The industries with the highest estimated rates of fatal traumatic injuries are mining (30.1 per 100,000 workers), construction (23.1), and agriculture (20.3).[22]

Occupational Exposure to Noise

Occupational exposure to noise is a widespread problem that has substantial impact on the prevalence of hearing loss among the working population. Recent estimates indicate that between 8 million and 10 million people work at sites where the level of noise—85 decibels or higher—presents an increased risk of noise-induced hearing loss.[18,23] One worker in four exposed occupationally to 90 decibels of noise over a working lifetime will develop noise-induced hearing impairment.

Recognition of Occupational Disease

Associations between occupational exposures and disease typically are recognized in three ways: through clinical observation, through epidemiologic analysis, and through toxicologic evaluation of chemical substances.

Clinical Recognition

The alert clinician is the key to clinical recognition of occupational disease.[7,24] Among the occupational illnesses that have been recognized by alert clinicians are angiosarcoma of the liver in workers exposed to vinyl chloride monomer,[6] lung cancer in workers manufacturing bischloromethylether,[25] and bladder cancer in aniline dye workers.[26]

The following are some the keys to the recognition of occupational illness:

- The clinician is alert to the possibility that any patient may have an occupational disease and therefore obtains at least a brief occupational history from every patient.
- The clinician possesses basic knowledge of the major occupational illnesses and of any particular occupational illnesses that are common in the area of practice.
- The clinician understands the basic concepts of epidemiology and toxicology and thus can consider new problems in a rational framework.
- The clinician is willing to expand the traditional medical approach to diagnosis and treatment by consulting with public health officials, industrial hygienists, and other specialists knowledgeable about the

work environment when he or she encounters a disease that may be related to work.

Although the clinician should be aware of the possibility of identifying a new occupational disease, it is even more important that he or she be able to recognize known occupational illnesses. The Sentinel Health Event (SHE) concept provides a useful mechanism to guide clinicians in the recognition of occupational diseases.[27] A SHE is a preventable disease, disability, or untimely death whose occurrence serves as a warning signal that the quality of preventive or therapeutic care may need to be improved. In 1983, to extend this concept to work-related disease, Rutstein et al.[12] developed a list of Sentinel Health Events (Occupational) (SHE(O)). The list provides information linking health conditions with industries, occupations, and possible toxic agents. By scanning the table, a clinician can identify pertinent industries or exposures in his or her practice area. An abbreviated list of SHE(O) is presented in table 2.

Epidemiologic Analysis

Epidemiologic recognition of a causal association between occupational exposure and human disease is illustrated by the studies of the link between benzene exposure and leukemia. Although this association was suspected clinically over 50 years ago, the causal relationship was not firmly established until epidemiologic studies were completed in the 1970s and the 1980s.[28,29] Another example in which a causal connection was established using epidemiologic tools is the recognition of the synergistic interaction between asbestos and cigarette smoking in the etiology of lung cancer.[30]

The major strategies used by epidemiologists to recognize occupational diseases are the cross-sectional medical study, the proportional mortality study, the historical cohort mortality study, and secondary analysis of vital statistics and other population-based health data. Cross-sectional studies are useful for the identification of acute, short-latency conditions or stable conditions that do not result in workers' leaving their jobs. (Workers who left work because of poor health are not included in these studies.) A limitation inherent in such studies is the

Table 2 Abbreviated List of Sentinel Health Events (Occupational). Adapted from reference 12.

Condition	Industry or occupation	Agent
Pulmonary tuberculosis	Physicians, medical personnel	*Mycobacterium tuberculosis*
Plague, tularemia, anthrax, rabies, other infections	Farmers, ranchers, hunters, veterinarians, laboratory workers	Various infections agents
Rubella	Medical personnel, intensive care personnel	Rubella virus
Hepatitis A, B, and Non-A, Non-B	Day care center staff, orphanage staff, medical personnel	Hepatitis A virus Hepatitis B virus
Ornithosis	Bird feeders, pet shop staff, poultry producers, veterinarians, zoo employees	*Chlamydia psittaci*
Malignant neoplasm of nasal cavities	Woodworkers, cabinet and furniture makers Radium chemists and processors Nickel smelting and refining	Hardwood dusts Radium Nickel
Malignant neoplasm of larynx	Asbestos industries and utilizers	Asbestos
Malignant neoplasm of trachea, bronchus, and lung	Asbestos industries and utilizers Topside coke ovenworkers Uranium and fluorspar miners Smelters, processors, users Mustard gas formulators Ion exchange resin makers, chemists	Asbestos Coke oven emissions Radon dauthers Chromates, nickel, arsenic Mustard gas Bis(chloromethyl) ether
Mesothelioma	Asbestos industries and utilizers	Asbestos
Malignant neoplasm of bone	Radium chemists and processors	Radium
Malignant neoplasm of scrotum	Automatic lathe operators, metal workers Coke oven workers, petroleum refiners	Mineral/cutting oils Soots and tars

Table 2 (continued)

Malignant neoplasm of bladder	Rubber and dye workers	Benzidine, naphthylamine, auramine, 4-nitrophenyl
Malignant neoplasm of kidney	Coke oven workers	Coke oven emissions
Lymphoid leukemia, acute	Radiologists, rubber industry	Ionizing radiation, unknown
Myeloid leukemia, acute	Occupations with exposure to benzene Radiologists	Benzene Ionizing radiation
Erythroleukemia	Occupations with exposure to benzene	Benzene
Hemolytic anemia, non-auto-immune	Whitewashing and leather industry Electrolytic processes	smelting Plastics industry
Aplastic anemia	Explosives manufacture Radiologists, radium chemists	TNT Ionizing radiation
Agranulocytosis or neutropenia	Explosives and pesticide industries Pesticides, pigments, pharmaceuticals	Phosphorus Inorganic arsenic
Toxic encephalitis	Battery, smelter, and foundry workers	Lead
Parkinson's disease (secondary)	Manganese processing, battery makers, welders	Manganese
Inflammatory and toxic neuropathy	Pesticides, pigments, pharmaceuticals Furniture refinishers, degreasing operations Plastics, rayon industries Explosives industry Battery, smelter, and foundry workers Dentists, chloralkali plants, battery makers Plastics industry, paper manufacturing	Arsenic and arsenic compounds Hexane MBK, CS_2, other solvents TNT Lead Mercury Acrylamide

Table 2 (continued)

Cataract	Microwave and radar technicians Radiologists Blacksmiths, glass blowers, bakers Moth repellant formulators, fumigators	Microwaves Ionizing radiation Infrared radiation Naphthalene
Noise effects on inner ear	Many industries	Excessive noise
Raynaud's phenomenon (secondary)	Lumberjacks, chain sawyers, grinders Vinyl chloride polymerization industry	Whole body, segmental vibration Vinyl chloride monomer
Extrinsic asthma	Jewelry, alloy and catalyst makers Polyurethane, adhesive, paint workers Plastic, dye, insecticide makers Foam workers, latex makers, biologists Bakers Woodworkers, furniture makers	Platinum Isocyanates Phthalic anhydride Formaldehyde Flour Red cedar and other wood dusts
Coalworkers' pneumo-coniosis	Coal miners	Coal dust
Asbestosis	Asbestos industries and utilizers	Asbestos
Silicosis	Quarrymen, sandblasters, silica processors, mining, metal, and ceramic industries	Silica
Talcosis	Talc processors	Talc
Chronic beryllium disease of the lung	Beryllium alloy workers, ceramic and cathode ray tube makers, nuclear reactor workers	Beryllium
Byssinosis	Cotton industry workers	Cotton, flax, hemp, and cotton-synthetic dusts

Table 2 (continued)

Acute bronchitis, pneumonitis, and pulmonary edema due to fumes and vapors	Alkali and bleach industries Silo fillers, arc welders Paper, refrigeration, oil industries Plastics industry	Chlorine Nitrogen oxides Sulfur dioxide Trimellitic anhydride
Toxic hepatitis	Solvent utilizers, dry cleaners, plastics industry Explosives and dye industries Fumigators, fire extinguisher formulators	Carbon tetrachloride, chloroform
Acute or chronic renal failure	Battery makers, plumbers, solderers Electrolytic processes, smelting Battery makers, jewelers, dentists Fire extinguisher makers Antifreeze manufacture	Inorganic lead Arsine Inorganic mercury Carbon tetrachloride Ethylene glycol
Infertility, male	Formulators DBCP producers, formulators, and applicators	Kepone Dibromochloropropane
Contact and allergic dermatitis	Leather tanning, poultry dressing plants, fish packing, adhesives and sealant industry, boat building and repair	Irritants (e.g. cutting oils, solvents, acids, alkalis; allergens)

difficulty of determining the temporal sequence of exposure and disease. A proportional mortality study compares the pattern of causes of death among a group of workers against that in a comparison population. It is relatively quick to perform, since it requires information about only those employees who have died. The historical cohort mortality study plays a preeminent role in occupational epidemiology because worker populations generally can be well defined through the use of employment records and seniority lists. The mortality experience of the cohort is determined by reviewing death certificates and other sources of information. Oftentimes, the limiting factor in cohort mortality studies is the poor quality of data on past exposures. Epidemiologic analysis of population-based health data can be useful for surveillance of large populations of workers and for the recognition of new exposure-disease associations. An example is the use of tumor registries to identify occupations with elevated risks of specific cancers. Case-control studies then can be performed to obtain detailed information from individuals about their past occupational exposures.

Toxicologic Evaluations

Toxicologic analysis of chemical substances is an important means of assessing causal relationships. The particular strength of toxicologic analysis as a tool for disease prevention derives from the fact that, ideally, it precedes occupational exposure. Thus, chemicals found in the laboratory to cause adverse health effects can be banned or strictly controlled, so as to minimize the exposure of workers.

Unfortunately, despite decades of concern about the toxic effects of chemical substances, little is known about the potential toxic effects of the majority of chemicals currently in commercial use. A study by the National Research Council[31] found that no information was available on the toxicity of approximately 80% of the 50,000 chemicals in commercial use (figure 1). Even for the groups of substances about which most is known—foods and drugs—reasonably complete information on untoward effects is available for only a minority of agents.

The Occupational History

The occupational history is the principal clinical instrument for the diagnosis of occupational disease.[32] Proper diagnosis of occupational disease is important because it facilitates the removal of the ill worker from continuing exposure, because it provides a guide to treatment, and because it fosters early identification and prevention of similar diseases in other workers.[12]

A systematic approach to taking an occupational history and to diagnosing occupational illness was developed by Goldman and Peters.[33] It proceeds through four stages:

(1) routine questions to assess possible occupational origins of illness in all patients

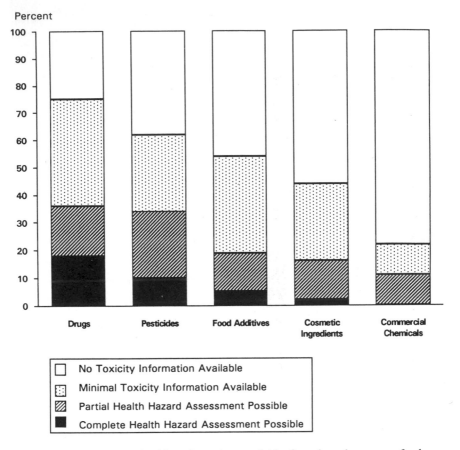

Figure 1 Toxicity and health information available for selected groups of substances. Adapted from reference 31.

(2) specific questions about any sources of exposure identified in stage 1

(3) identification of any hazardous materials reported in stages 1 and 2 and their uses

(4) followup consultation and resolution of any problems identified above.

It may not be possible to obtain a detailed occupational history on every patient. However, the clinician should routinely ask screening questions of every patient that provide an indication as to whether a complete occupational history is warranted. At a minimum, every patient should be asked about his or her current job by industry and occupation, and about the longest-held previous jobs. A general question should be asked about occupational exposures to chemicals, fumes, gases, dust, noise, radiation, or other physical hazards at work. If the patient reports exposure to any of these agents, it may be useful to ask if he or she thinks there is a health hazard at work. In addition to these screening questions, the clinician should pay attention during the medical history and review of systems to any temporal relationships reported between work and the onset of symptoms. Any history of suspicious exposures requires that the clinician obtain a complete occupational history. A complete occupational history also is appropriate for evaluation of any of the conditions listed as a Sentinel Health Event (Occupational) in table 2. A complete occupational history will obtain the following data:

- descriptions of *all* jobs ever held
- detailed information about work exposures in each job, including working conditions and use of personal protective equipment.
- information on the time course of the patient's symptoms and their temporal relation to work.
- information on significant symptoms of illness among co-workers.
- information on non-work exposures and other possibly relevant factors, such as exposures due to hobbies.

Surveillance of Occupational Disease

Surveillance is a technique for systematic recognition of occupational diseases and hazards.[34] Hazard surveillance provides a means for assessing toxic occupational exposures in a population and thus for assessing risk.[35]. Occupational disease surveillance provides a means of assessing the amounts and types of occupational disease, the time trends, and the distribution according to geography, industry and occupation. Hazard surveillance should identify the chemicals in use, the industries and occupations in which they are used, and the extent and magnitude of worker exposure. These two types of surveillance complement each other. Each is an integral component of a complete occupational health surveillance system.

Occupational disease surveillance is based primarily on legal requirements that employers maintain records of occupational injuries and illnesses. These records serve as the basis for the Bureau of Labor Statistics' Annual Survey. It is widely recognized that this system leads to a substantial underestimation of the true extent of occupational illness.[36] Another widely cited data source for occupational disease surveillance is a compilation of physicians' report to workers' compensation boards. However, workers' compensation was not designed for the purpose of surveillance. Like the employer-based system, this data source underestimates the extent of occupational illnesses.

Another approach adopted recently by some states is the establishment of requirements that medical laboratories report to the state health department all cases of certain abnormal laboratory results. Most notably, this approach has been used to monitor increased absorption of heavy metals. Public health officials then follow up on reported cases to determine whether they suggest the presence of an occupational hazard. Examples of such surveillance programs are the New York and California Heavy Metals Registries.[37,38]

A fundamental limitation of these surveillance systems is that they rely on reporting mechanisms in which recognition of the problem occurs only after the fact. Primary and secondary prevention of occupational disease require a more direct strategy in which surveillance is conducted in the workplace. Disease surveillance in the workplace—consisting of biological monitoring and medical screening of workers—uses the health history and results of periodic physical and laboratory examinations to estimate levels of exposure to toxins and to assess early effects of exposure. Biological monitoring techniques are available to assess body concentrations of a number of toxic substances.[39] Medical screening of apparently well workers may be used to detect workers at risk prior to the onset of disease or dysfunction (primary prevention) or to detect workers at an early stage of disease, when effects are still potentially reversible (secondary prevention). The criteria for the implementation of medical screening programs in occupational medicine are similar to those for other health screening programs, and are based on estimation of the validity, sensitivity, specificity, predictive value, cost effectiveness, and acceptability of the screening procedure.

Prevention of Occupational Disease

Primary prevention of occupational disease requires elimination or reduction of hazardous exposures. Secondary prevention depends on the ability to effectively identify potential work-related illness through screening of workers at high risk. Tertiary prevention—the prevention of complications and disability resulting from existing illness—depends on the development and application of appropriate diagnostic techniques for identification of persons with established occupational illness. Prevention on all three levels requires solid information on the potential effects of specific occupational exposures, as well as data on the industries and occupations in which hazardous substances are used. The most effective prevention strategy is the primary prevention of exposure. Reductions in exposure may be accomplished by the following techniques, listed in descending order of preference:

- Substituting a less hazardous material is the most efficacious method of controlling a workplace hazard. Substitution of a less hazardous process or equipment also represents a meaningful control strategy. For example, substitution of a continuous process for an intermittent process almost always results in a decrease of exposure.
- Engineering controls—such as ventilation, process isolation, or enclosure—may be used to reduce worker exposure to toxic substances. Ventilation is one of the most effective and widely used control measures.
- Alteration of work practices can help to reduce exposure to hazards. An example is wet-sweeping asbestos dust rather than dry-sweeping.
- Administrative controls are methods of controlling total worker exposure by job rotation, work assignment, or time periods spent away from the hazard. With administrative controls, the level of exposure to the hazard is not diminished; instead, the duration of individual exposure is reduced and exposure is spread more widely among the workforce. The most common use of administrative controls is to reduce overall noise exposure through rotation.
- Programs for encouraging personal hygiene constitute another approach to reducing exposure. In some instances the employer may encourage or require showers and a change to clean clothes at the

end of the workday. Naturally, the employer should provide showers, changing facilities, lockers, and work clothes if such measures are indicated.

- Use of personal protective equipment, such as respirators, gloves, protective clothing, and ear plugs or muffs, can play an important role if carefully designed programs are in place and if the equipment is checked regularly. It is important, however, to recognize that programs of personal protection never constitute as efficient or acceptable a means of protection as engineering or process controls.

Occupational Disease in the Future

Major changes are occurring in the U.S. workforce. The principal shift is away from a manufacturing-based economy to an economy that is based on the provision of services. In consequence, the hazards of exposure in heavy industry (such as coal workers' pneumoconiosis or silicosis) may be less important in the future, and the hazards associated with office work (such as video display terminals or exposures to new synthetic materials and solvents such as those used in the electronics industry) may become more important.

A major future need in occupational health will be for the much more widespread use of pre-market testing of chemical substances. The widespread lack of toxicity data on chemicals in commerce and the current inadequacies in pre-market evaluation of the toxicity of newly developed chemical compounds will need to be corrected.[31]

Increased research in the years ahead will be directed toward the development of biological markers of both exposure and disease.[39] Biological markers of exposure, such as the blood lead level, provide a means for individualized assessment of exposure. Biological markers of disease or of subclinical dysfunction are intended to detect early, pre-clinical pathophysiologic changes caused by toxic agents. The use of such markers in worker screening and in population studies may be expected to permit earlier identification of adverse effects when they are at an early and still potentially reversible stage.

Finally, the role of the primary-care physician in the recognition and management of occupational diseases must be increased dramatically. The American Board of Medical Specialties identifies only slightly more than 1,000 physicians in the United States today who are board-certified in occupational medicine. Of the more than 104 million individuals in the workforce, approximately 70% work in facilities with no medical services. These numbers demonstrate that there is no alternative for working men and women but to rely on their primary-care physicians for the recognition and management of potential work-related disorders that may be attributable to hazardous exposures encountered at work.[40] A report of the Institute of Medicine concluded that "at a minimum, all primary care physicians should be able to identify possible occupationally or environmentally induced conditions and make the appropriate referrals for followup."[7] The American College of Physicians has stated the following:

Physicians, especially primary care internists, must address the occupational health care needs of their patients. Physician responsibilities in providing care to patients of working age include identification of occupational and other environmental health risks, treatment of disease and injury, and patient counseling about preventive behavior.

Physicians have a responsibility to improve the health of the population by working to prevent occupational and other environmental risks that cause injuries and diseases.[11]

References

1. Landrigan, P. J. The Burden of Occupational Disease in the United States. Testimony before the Committee on Labor and Human Resources, United States Senate, Oversight Hearings on the Occupational Safety and Health Act, April 18, 1988.

2. Baker, E. L., et al. Lead poisoning in children of lead workers: Home contamination with industrial dust. *New England Journal of Medicine* 296 (1977): 260–261.

3. Chisolm, J. J., Jr. Fouling one's own nest. *Pediatrics* 62 (1978): 614–617.

4. Selikoff, I. J., Churg, J., and Hammond, E. C. Asbestos exposure and Neoplasia. *Journal of the American Medical Association* 188 (1964): 22–26.

5. Landrigan, P. J. Toxic exposures and psychiatric disease—lessons from the epidemiology of cancer. *Acta Psychiatrica Scandinavica* 67 (1983), suppl 303: 6–15.

6. Creech, J. L., Jr., and Johnson, M. N. Angiosarcoma of the liver in the manufacture of polyvinyl chloride. *Journal of Occupational Medicine* 16 (1974): 150–151.

7. Institute of Medicine. *Role of the Primary Care Physician in Occupational and Environmental Medicine.* National Academy Press, 1988.

8. Levy, B. S. The teaching of occupational health in United States medical schools. Five-year follow-up of an initial survey. *American Journal of Public Health* 75 (1975): 79–80.

9. Markowitz, S., and Landrigan, P. J. The magnitude of the occupational disease problem: An investigation in New York State. *Toxicology and Industrial Health* 5 (1989): 9–30.

10. Nicholson, W. J., Perkel, G., and Selikoff, I. J. Occupational exposure to asbestos: Population at risk and protected mortality—1980–2030. *American Journal of Industrial Medicine* 3 (1982): 259–311.

11. *The Role of the Internist in Occupational Medicine.* American College of Physicians, 1984.

12. Rutstein, D. D., et al. Sentinel health events (occupational): A basis for physician recognition and public health surveillance. *American Journal of Public Health* 73 (1983): 1054–1062.

13. President's Report on Occupational Safety and Health. Government Printing Office, 1972.

14. Levy, B. S., and Wegman, D. H., eds. *Occupational Health—Recognizing and Preventing Work-Related Diseases,* second edition. Little, Brown, 1988.

15. Rom, W. N., ed. *Environmental and Occupational Medicine,* second edition. Little, Brown, 1992.

16. Zenz, C. *Occupational Medicine: Principles and Practical Application,* second edition. Year Book Medical Publications, 1988.

17. National Institute for Occupational Safety and Health. *Proposed National Strategies for the Prevention of Leading Work-Related Diseases and Injuries—Part 1.* Association of Schools of Public Health, 1986.

18. National Institute for Occupational Safety and Health. *Proposed National Strategies for the Prevention of Leading Work-Related Diseases and Injuries—Part 2.* Association of Schools of Public Health, 1988.

19. National Institute for Occupational Safety and Health. *National Occupational Hazard Survey 1972–1974.* Publication 78–114, Department of Health, Education and Welfare, 1978.

20. Infante, P. F., and Legator, M. S., eds. *Proceedings of a Workshop on Methodology for Assessing Reproductive Hazards in the Workplace.* National Institute for Occupational Safety and Health, 1980.

21. Scharden, J. L. *Chemically Induced Birth Defects.* Marcel Dekker, 1985.

22. Traumatic occupational fatalities—United States, 1980–1984. *Morbidity and Mortality Weekly Report* 36 (1978): 461–470.

23. Self-reported hearing loss among workers potentially exposed to industrial noise—United States. *Morbidity and Mortality Weekly Report* 37 (1988): 158–167.

24. Miller, R. W. The discovery of human teratogens, carcinogens and mutagens. Lessons for the Future. In *Chemical Mutagens,* ed. A. Hollander and F. W. deSevres. Plenum, 1978.

25. Figueroa, W. G., Raszowski, R., and Weiss, W. Lung cancer in chloromethyl methyl ether workers. *New England Journal of Medicine* 228 (1973): 1096–1097.

26. Rehn, L. Blasengeschuwuelste bei Fuchssinarbeitern. *Archiv für klinische Chirurgie* 50 (1895): 588–600.

27. Rutstein, D. D., et al. Measuring the quality of medical care—A clinical method. *New England Journal of Medicine* 294 (1976): 582–588.

28. Infante, P. F., et al. Leukaemia in benzene workers. *Lancet* 2 (1977): 76–78.

29. Rinsky, R. A., et al. Benzene and leukemia—an epidemiologic risk assessment. *New England Journal of Medicine* 316 (1987): 1044–1050

30. Selikoff, I. J., Hammond, E. C., and Churg, J. Asbestos exposure, smoking and neoplasia. *Journal of the American Medical Association* 204 (1968): 106–112.

31. National Research Council. *Toxicity Testing—Strategies to Determine Needs and Priorities.* National Academy Press, 1984

32. Hamilton, A. *Exploring the Dangerous Trades.* Little, Brown, 1943.

33. Goldman, R. H., and Peters, J. M. The occupational and environmental health history. *Journal of the American Medical Association* 246 (1981): 2831–2857.

34. Langmuir, A. D. The surveillance of communicable diseases of national importance. *New England Journal of Medicine* 268 (1963): 182–192.

35. Wegman, D. H., and Froines, J. R. Surveillance needs for occupational health. *American Journal of Public Health* 75 (1985): 1259–1261.

36. National Research Council. *Counting Injuries and Illnesses in the Workplace. Proposals for a Better System.* National Academy Press, 1987.

37. Maizlish, N., et al. Surveillance of blood lead in California adults, 1987. Presented at Annual Meeting of American Public Health Association, Boston, 1988.

38. Stone, R. R., Marshall, E. G., and Marion, D. R. Patterns of lead exposure in New York State. Abstract, Annual Meeting of American Public Health Association, Boston, 1988.

39. Goldstein, B., et al. Biological markers in environmental health research. *Environmental Health Perspectives* 74 (1987): 3–9.

40. Rosenstock, L. Occupational medicine: Too long neglected. *Annals of Internal Medicine* 95 (1981) 664–676.

Radiation and Health: Nuclear Weapons and Nuclear Power

Kenneth Lichtenstein, M.D.
Ira Helfand, M.D.

At least three major nuclear reactor accidents have occurred since the advent of nuclear power: one at Windscale (now Sellafield), England, in 1957, one at the Three Mile Island plant in Pennsylvania in 1979, and one at Chernobyl in the former Soviet Union in 1986. Hundreds of other "significant" radiation accidents have occurred at military, research, and commercial facilities.[1] After each accident, scientists rushed to answer the public's urgent question: How many cancers might result in the long run from the release of radioactivity?

The experts' estimates vary widely. For example, after the Chernobyl explosion U.S. Department of Energy scientists predicted 28,000 "excess" cancers—cases in excess of the number normally expected—in Europe and the western Soviet Union.[2] John Gofman of Lawrence Livermore National Laboratory argued that such estimates were based on far too optimistic assumptions; the cancer risk, he claimed, may be 30 times higher than officially adopted estimates.[3] While there remains significant uncertainty about these projections, most studies have estimated the cancer mortality from Chernobyl to be in the range of 17,000–60,000.[4-7]

Adding to the public confusion about radiation danger, medical research has recently produced new evidence that low-level radiation carries health hazards. Radiation protection standards are being revised accordingly. The public is left wondering why the relation between cancer and low-level radiation is so controversial—and which authorities to believe.

The biological science of how radiation affects the body is well understood, but the epidemiological science that tries to relate very low radiation exposures to subsequent cancer and other illness is based on still unproven—some say unprovable—hypotheses. This lack of scientific certainty fuels the public controversy over what level of radiation exposure is acceptable.

But public perceptions of the risk of nuclear technology depend only in part on scientific evidence. Kai Erikson's interviews with nearby residents after the accident at Three Mile Island document the special dread that radiation inspires.[8] Many radiation scientists have long complained that people do not understand that smoking and driving are consequently much more dangerous than exposure to low levels of radiation. And some blame the media for misrepresenting the Three Mile Island accident, in which the officially reported release of radioactive iodine (I-131) was quite small (13–20 curies, versus 50 million curies at Chernobyl).[1] Erikson points out that, for the public, radiation's invisibility and its lingering effects make it unlike other dangers. Public policy must take this special fear into account.

For many people the dangers are real. Radioactive waste has polluted large areas in and near nuclear weapons production plants across the United States and the former Soviet Union. For example, the health of residents in Washington and Idaho was endangered during the 1940s and 1950s by deliberate releases of radiation from the Hanford nuclear weapons plant. A comprehensive effort is underway to reconstruct the radiation doses that communities around and downwind of the Hanford plant have received over the decades.

The Biology of Radiation

Ionizing radiation is given off by decaying radioisotopes, or radionuclides, typically as alpha or beta particles or gamma rays. Ionizing radiation can penetrate a human cell and then ionize (i.e., knock an electron loose from) a cellular chemical, altering molecules important for normal functioning. When this happens (and it happens frequently, because we live with natural or "background" radiation—see figure 1), cells usually

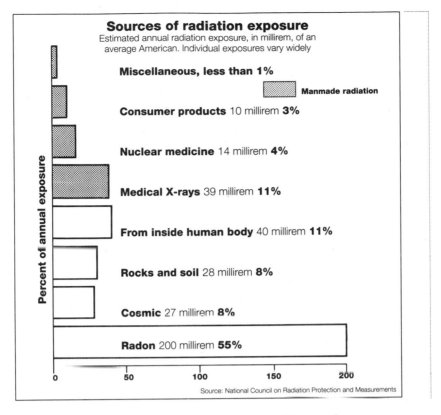

Sources of radiation exposure

Estimated annual radiation exposure, in millirem, of an average American. Individual exposures vary widely

Percent of annual exposure

Miscellaneous, less than 1%

Manmade radiation

Consumer products 10 millirem **3%**

Nuclear medicine 14 millirem **4%**

Medical X-rays 39 millirem **11%**

From inside human body 40 millirem **11%**

Rocks and soil 28 millirem **8%**

Cosmic 27 millirem **8%**

Radon 200 millirem **55%**

0   50   100   150   200

Source: National Council on Radiation Protection and Measurements

Figure 1   Estimated annual radiation exposure, in millirems, of an average American. (Individual exposures vary widely.) Source: National Council on Radiation Protection and Measurement.

repair themselves. But occasionally they die or are transformed. Other electromagnetic radiation, such as microwaves or light, may also injure cells but does not cause ionization.

Ionizing radiation produces two kinds of cell injury, one immediate and the other delayed. High-level doses—a term used loosely for exposures over 100 rem—inevitably produce the kind of immediate, direct effects seen in the Chernobyl firefighters: skin burns, hair loss, bone marrow destruction, and damage to the intestinal lining. The severity of the Chernobyl workers' injuries, some of which were fatal, was directly related to the amount of radiation received. But the effects of low doses—a loose term for exposures under 10 rem—are not immediately visible and are far less predictable, because they involve the cancerous transformation of cells. The complex process that produces cancerous tumors has many intermediate steps, and the changes may happen in some persons and not in others. It may be years before epidemiologists, using statistical methods, note an increased incidence of cancer in the population of exposed persons (for example, uranium miners who received low doses of radiation by inhaling radon gas and other radioisotopes).[9,63]

Most, but not all, scientists believe that such stochastic or probabilistic effects of radiation are also directly related to the radiation dose, and that they can occur at any dose, no matter how small. This is called the *linear, no-threshold hypothesis;* it means that all exposure to radiation presents some risk to human health.

That hypothesis is still controversial. Until recently, most radiation scientists felt that, because cells repaired themselves, no association with cancer would be found at radiation doses below a certain threshold. This was the majority view of the National Academy of Sciences committee that, in 1980, produced its third report on the biological effects of ionizing radiation (BEIR III). But BEIR V, published in 1990, unanimously accepts the linear, no-threshold hypothesis—a significant change in scientific opinion.[10] A few scientists, including John Gofman,[3] believe there is evidence that low doses produce more cancer per unit of radiation than high doses—the *supralinear hypothesis.* Others believe that, because there are no definitive data on low doses, any hypothesis is guesswork.[11]

Although the relation between very low doses of radiation and cancer is still unresolved, there is little argument about the effects of high and moderately high doses of radiation.[1,12] The direct, immediate effects of radiation are seen in all cases, from the highest doses down to about 100 rem. Scientists also agree that persons exposed to levels of 10–50 rem have an increased likelihood of developing cancer. These conclusions are based on experiences with x rays, as well as on observations of bomb and accident victims. Early radiologists, unaware of the potential harm from the new Roentgen rays (x rays), suffered both direct effects (skin burns and hair loss) and indirect effects (higher rates of leukemia some years later). In Great Britain and the United States in the 1930s, several thousand children were treated for scalp ringworm with high doses (200–400 rem) of x rays. Other patients received local irradiation for spinal disease and benign gynecological disorders, including breast inflammation. All these groups later were shown to have increased rates of leukemia and certain types of cancer.[13]

The Epidemiologists' Problem

Epidemiologists investigate the causes of disease in groups rather than in individual patients. Proving causation is the most difficult part of their task, and it relies on judgment. Low-level radiation is associated with increased cancer incidence, but association is not causation. Scientists test associations for strength, consistency, specificity, and plausibility before making judgments about causation. For example, lung cancer is more common in smokers than in non-smokers, and the consistent judgment of most scientists, after the usual tests, is that smoking "causes" cancer.

Although cancer is clearly associated with radiation at higher doses, extrapolating such association to lower doses is difficult because mortality rates from cancer are influenced by a large number of factors. Age is the most important. For example, Florida has an annual cancer mortality rate of about 250 per 100,000 population, while Alaska's is about 80 per 100,000. The difference is due primarily to the age difference of the two states' residents. Sex, race, urban or rural residence, occupation, diet, and lifestyle also enter in, and adjustments must be made for each of these factors.

Cancer causes about 20% of all deaths in the United States. Because radiation-induced cancers are not distinguishable from those caused by other agents or from those that occur "naturally," the only way to associate cancer with radiation is to compare a group of irradiated persons with an identical group of unirradiated persons. Alice Stewart, a physician-epidemiologist and a central figure in the controversy over low-level radiation and cancer, has argued for years that scientists who claim that low-level radiation has no measurable effect have ignored the very great difficulties in making the two groups studied identical.[14,15] (The technical term for this difficulty is *selection bias*.)

And because cancer is common, one must study very large numbers of persons in order to explore specific causes—otherwise the inci-

dence of cancer in an irradiated group would not be measurably different from the natural incidence. But there are very few large populations
who have been exposed to known levels of radiation. The 106,000
surviving victims of Hiroshima and Nagasaki are the best known and
studied. The doses they received have been painstakingly reconstructed
as part of the work of a large, ongoing research program set up in 1946:
the Atomic Bomb Casualty Commission, now known as the Radiation
Effects Research Foundation (RERF), based in Hiroshima.[11]

Ironically, the tragic accident at Chernobyl in April 1986 may
provide another opportunity for a definitive, long-term, forward-look-
ing (prospective) study of low-level radiation exposure. At a Vienna
meeting of the International Atomic Energy Agency in 1986, Soviet
scientists promised to organize such a followup study of the hundreds
of thousands of persons exposed.[16] An authoritative and fully documented account of the Soviet management of the accident and its
impacts on health, environment, economics, agriculture, and nuclear
power has been published by Zhores Medvedev, a former Soviet scientist who now lives and works in London.[17]

Two additional groups of workers exposed to radiation in the
nuclear weapons production program of the former Soviet Union have
been described in recent publications. Workers and residents at the
Chelyabinsk nuclear weapons production facility were exposed in the
facility, from radioactive waste poured directly into the Techa River and
after a massive explosion. Three physicians from Chelyabinsk have recently published a unique report of their decades-long studies of human
radiation exposures in their region.[62]

In southern regions of the former East Germany, uranium miners
and metal workers in the Soviet bomb program had prolonged high-
level occupational radon and other nuclide exposures. It turns out that
longitudinal records of their health status, and in many cases radiation
exposures, were kept by area physicians.[63] These two populations may
represent tragic but important additional cohorts for the study of radiation's health effects.

The only large group of Americans exposed to known doses of
low-level radiation are the workers employed by the Department of
Energy and its contractors at nuclear weapons production facilities. The
Argonne National Laboratory estimates that at least 600,000 persons,

---

and perhaps as many as a million, have worked in these facilities at one time or another.[18]

New Information

BEIR V raised the estimate of cancer risk associated with low-level radiation exposure to 2–4 times the estimate used in BEIR III. In the intervening decade, RERF scientists observed more cancer deaths among the bomb victims than had been predicted. At the same time, improved calculations lowered the estimates of the gamma radiation released by the bombs. In other words, less radiation had produced more cancer. And, as was mentioned above, investigators have abandoned the notion of a threshold below which no cancer effect would be observed.

In response to these findings, the International Commission on Radiological Protection is recommending reducing the exposure allowed for nuclear workers, medical personnel, and others in contact with radiation from 5 rem per year to 2. As Catherine Caulfield points out in her book *Multiple Exposures*,[13] the standards for occupational and public radiation exposure have become progressively more stringent over the last 60 years (see table 1). In 1987 new information on radon gas in households led the National Council on Radiation Protection and Measurement to more than double its estimate of the average American's background radiation exposure, from 170 millirem per year to 360 millirem (see figure 1). That is equivalent to about 18 chest x-ray film exposures, and it is still small relative to the 5-rem annual exposure which the federal government allows in the workplace.

New research findings appear to indicate that low levels of radiation produce genetic damage that can be transmitted to offspring. Gardner et al. report that children of men exposed to low-level radiation may have a higher risk of leukemia.[19] The results are from a British National Research Council study of workers at the Sellafield nuclear reprocessing plant. This finding is surprising, because no evidence of increased leukemia or cancer was found in 75,000 children whose parents had survived the bombing of Hiroshima or Nagasaki. Although scientific reviewers believe the Sellafield study was performed well, some scientists have disagreed with Gardner's conclusion that the chil-

---

Table 1 Changing radiation standards: U.S. recommended maximum permissible whole-body doses of external radiation (above background).

Year	Allowed exposure (rems/year)
Occupational exposure	
1934	30
1949	15
1957	5
1960	5
1977	5
1987	5
1990*	2
General public exposure	
1956	0.50
1960	0.17
1987	0.10

*proposed

dren's leukemia was due to radiation damage to the father's sperm.[20] Chemical agents or latent viruses, they have suggested, might also have explained the findings.

The Environmental and Health Effects of Nuclear Weapons

Nuclear weapons production and testing, conducted under a curtain of secrecy for nearly 50 years without the benefit of proper scientific or medical scrutiny, has resulted in significant contamination of the planet and potentially disastrous human health consequences. Radionuclides have been dispersed worldwide, while worker safety and public health have been largely ignored. The nuclear industry and the governments involved have chosen to minimize the significance of this radioactive contamination by including it with normal "background" radiation (see figure 1).

As a result of these policies, all living organisms, including humans, have incorporated plutonium (Pu) and other man-made radionuclides with carcinogenic, teratogenic, and mutagenic properties into their tis-

Figure 2 Department of Energy nuclear weapons facilities. Source: Office of Technology Assessment (reference 23).

sues. Some of these substances will remain in the cells of living organisms for very long periods of time (one isotope of plutonium, Pu–239, has a half-life of approximately 24,000 years).

The effects of these activities on human health have yet to be fully described and understood. Relevant data have not been available for independent research, owing to a stated need for national security, and those health studies performed internally by the nuclear weapons industry have been done without peer review or confirmatory independent analysis. Much of this research has been flawed methodologically, and unfavorable outcomes have been suppressed.[21]

It is often difficult to determine the effects of environmental degradation on human health, as the science of risk assessment used to calculate such effects is imprecise. It is particularly so at nuclear weapons facilities. Little is known at these sites about the degree or duration of exposure of workers and surrounding populations to radionuclides and hazardous chemicals.[22] Efforts are only now underway in the United States to begin to reconstruct exposure levels for workers and for human populations adjacent to these facilities.[15] Furthermore, pre-exposure health records are, with few exceptions, not known. The health consequences of man-made global radioactive contamination are, as a result, poorly understood.

Many examples of either intentional or accidental releases of radionuclides or hazardous chemicals into the environment from nuclear weapons production and testing facilities in different countries are known. These sites are among the most polluted areas on Earth.

United States

Fernald

This facility, located 20 miles northwest of Cincinnati, produces uranium metal for nuclear weapons. It is a 1,450-acre site surrounded by rural farmland. Contaminated groundwater poses the greatest threat to human health; private, community, and industrial drinking-water wells are all contaminated. In addition, surface water and fresh water sediment contamination have been confirmed both on and off site. Soil on the site is also contaminated. Major hazardous substances released into the environment include a number of radionuclides, heavy metals, inorganic chemicals, volatile organic compounds (VOCs), and asbestos. Since 1951, anywhere from 394,000 to 3,100,000 pounds of uranium dust were intentionally released into the air around the facility.[23]

Hanford

The Hanford Reservation encompasses 360,000 acres of the Columbia River Basin in southeastern Washington. This facility produces weapons-grade plutonium. Groundwater contamination is widespread, with tritium and nitrates found in plumes totaling 122 square miles. Carbon tetrachloride, chromium, cyanide, trichlorethylene, uranium, cobalt (Co-60), technetium (Tc-99), iodine (I-129), and strontium (SR-90) have also been found in local groundwater at levels that exceed current drinking-water standards. Wastewater containing hazardous and radioactive constituents is still being discharged into the soil, and the U.S. Department of Energy plans to continue this practice until June 1995. Surface-water sediment is contaminated as a result of past disposal practices. The extent of soil contamination by hazardous chemicals is unknown. Radionuclides, on the other hand, have been measured, and uptake by vegetation has been documented.[23]

The most dramatic incident here, which demonstrated a profound indifference to the potential effects of radioactive substances on human life, occurred during an experiment on December 2, 1949, when scientists deliberately released over 20,000 curies of xenon (Xe-133) and 7,780 curies of iodine (I-131). The surrounding population was neither evacuated nor informed, and no followup health studies were undertaken.[24] Currently, the Centers for Disease Control is funding a study of thyroid cancer in the exposed population.[25]

Idaho National Engineering Laboratory
This facility, which reprocesses reactor fuel to recover uranium (U-235), covers 570,000 acres in southeastern Idaho. It is located 22 miles from the city of Idaho Falls. Groundwater appears to be the main contaminated medium. It contains carbon tetrachloride, tricholorethylene, I-129, Pu-238, Pu-239, Sr-90, and tritium. This site has not been well studied; further efforts to do so are underway.[23]

Nevada Test Site
The Nevada Test Site is located 65 miles northwest of Las Vegas. It occupies 1,350 square miles. To date the United States has conducted 100 atmospheric and 714 underground nuclear tests at this location.

Despite the many years of operation, environmental contamination has been assessed only in a preliminary fashion. The groundwater has been documented to contain krypton (KR-85), chlorine (C1-36), ruthenium (Ru-106), Tc-99, and I-129. Current studies have indicated that it has not migrated off the site.[26] However, numerous studies have been done on downwind communities. Stevens et al. found "a weak association between bone marrow dose and all types of leukemia" in those living downwind for all ages and all time periods after the exposure.[27] The relative risk of leukemia for all ages was 1.72 times the normal risk, and the relative risk of acute leukemia for ages 0–19 was 7.82 times the normal risk.

The DOE sponsored a dose-reconstruction study of fallout from the NTS. The total gamma exposure was estimated at 86,000 person-rems, including 320 person-rems from ventings of underground tests from 1963 through 1975. The estimated population that received this total dose is 180,000. The average external gamma dose per person, therefore, is about 480 millirem.[28]

Oak Ridge Reservation

This facility covers 58,000 acres in eastern Tennessee, 15 miles from Knoxville. Among its many missions, it produces nuclear weapons components. Groundwater, surface water, and soil are contaminated by a large number of radionuclides, heavy metals, and VOCs. The extent of this contamination is still under investigation. Sediment of the Clinch River and several of its tributaries is contaminated with cesium (Cs-137), Pu-239, and Pu-240.[23]

Rocky Flats Plant

The Rocky Flats plant is 16 miles from downtown Denver, and some of Denver's suburbs are adjacent to it. Two creeks run through the property and into the Great Western Reservoir and Standley Lake, both public drinking-water reservoirs. The sediment of both reservoirs is contaminated with plutonium (Pu-239, Pu-240) and with tritium; the latter was accidentally released into the Great Western Reservoir in 1973. Tritium was detected in the urine of residents drinking water from this reservoir for 3 years after the release.[29]

Rocky Flats is a metal foundry that manufactures the plutonium "triggers" for new warheads and extracts plutonium from retired ones. Large amounts of solvents are used in these processes. More than 7,500 chemicals have been identified at the plant site.[30]

Besides ongoing plutonium emissions from daily operations, there have been three accidents that have released plutonium into the air over the Denver metropolitan area (population: 1.7 million). The first was an explosion and fire in 1957.[31] The second involved windstorms that dispersed soil contaminated with radioactive lathe oil that had leaked from corroded 55-gallon storage drums. The third incident was a smaller fire that breached the building in which it had started.[32]

Soil, sediment, groundwater, and surface water have all been contaminated with a large number of radionuclides, heavy metals, inorganic chemicals, VOCs, PCBs, and asbestos. Sediment, soil, and surface-water contamination off the site have been confirmed. The characterization of off-site groundwater is not yet complete.[23]

Several epidemiologic studies have been performed on the adjacent population. Some have suggested a higher mortality from cancer in people living closer to the plant.[31] Others have refuted these findings.[33] Wilkinson et al. studied the workers and found a higher mortality rate from hematopoietic neoplasms in those workers whose internal body burden of plutonium exceeded 2 nanocuries.[34]

Savannah River Plant

This facility produces tritium and plutonium for nuclear weapons. It covers 192,000 acres on the north bank of the Savannah River, 13 miles south of Aiken, South Carolina. Groundwater contamination with multiple radionuclides, inorganic chemicals, and VOCs has been found on site only, although studies are incomplete. Sediment, soil, and surface water, likewise, are only now under review.[23]

The above findings are summarized in table 1 and in figure 3.

Worldwide Nuclear Testing

Owing to the secrecy associated with nuclear weapons production and testing, little has been published about these facilities. However the International Physicians for the Prevention of Nuclear War (IPPNW) and the Institute for Energy and Environmental Research recently completed a review of the worldwide health and environmental effects of nuclear weapons testing.[35] They estimate the global inventory of long-lived radionuclides as a result of the 423 atmospheric tests conducted from 1945 to 1980 to be 11–13 million curies of S-90, 17–21 million curies of Cs-137, 10 million curies of carbon-14, and 225,000 curies of Pu-239 (4,200 kg). These radionuclides have been dispersed around-the world, with the majority of the fallout in the northern hemisphere.

The approximately 1,400 underground nuclear tests conducted around the world from 1957 through 1989 took place at nine sites in

Table 2 Radioactive and toxic contamination at major U.S. nuclear weapons production facilities, 1990. Sources: "Status of major nuclear weapons production facilities: 1990," *PSR Monitor,* September 1990; R. Alvarez and A. Makhijani, "Hidden legacy of the arms race: Radioactive waste," *Technology Review,* August–September 1989; other sources. (Source of table: reference 21.)

Facility and task	Observation
Feed materials Production Center, Fernald, Ohio (converts uranium into metal ingots)	Since plant's opening, at least 205 tons of uranium oxide (and perhaps 6 times as much as that) released into the air. Off-site surface and ground-water contaminated with uranium, cesium, thorium. High levels of radon gas emitted.
Hanford Reservation, Wash. (recycles uranium and extracts plutonium)	Since 1944, 760 billion liters of contaminated water (enough to create a 12-meter-deep lake the size of Manhattan) have entered groundwater and Columbia River; 4.5 million liters of high-level radioactive waste leaked from underground tanks. Officials knowingly and sometimes deliberately exposed the public to large amounts of airborne radiation in 1943–56.
Savannah River, S.C. (produces plutonium and tritium)	Radioactive substances and chemicals found in the Tuscaloosa aquifer at levels 400 times greater than government considers safe. Released millions of curies of tritium gas into atmosphere since 1954.
Rocky Flats, Colo. (assembles plutonium triggers)	Since 1952, 200 fires have contaminated the Denver region with unknown amount of plutonium. Strontium, cesium, and cancer-causing chemicals leaked into underground water.
Oak Ridge Reservation, Tenn. (produces lithium-deuteride and highly enriched uranium)	Since 1943, thousands of pounds of uranium emitted into atmosphere. Radioactive and hazardous wastes have severely polluted local streams flowing into the Clinch River. Watts Bar Reservoir, a recreational lake, is contaminated with at least 175,000 tons of mercury and cesium.

the United States, over 50 in the USSR, nine in the Pacific, two in Algeria, one in China, one in India, and possibly one in the South Atlantic. The underground inventory of radionuclides due to these tests consists of 5.3 million curies of SR-90, 8.4 million curies of CS-137, and 0.2 million curies of Pu-329.[23]

The South Pacific

The United States conducted 106 tests in the Pacific, mostly in the Marshall Islands, an island group with a population of 36,000 inhabitants. The British conducted nine tests at Johnston Atoll and Christmas Island, and the French have conducted 167 (44 atmospheric) at Moruroa and Fangataufa in French Polynesia.

In the Marshall Islands the most seriously affected atolls were evacuated and resettled several times. After "Bravo" (the code name for the test of a 15-megaton thermonuclear bomb on March 14, 1954), the exposure rate of the population on Rongelap Island was recorded at 1.2–2.3 roentgens per hour. Eighty-six people received an estimated whole-body exposure averaging 190 rem. Ten percent of the people experienced acute radiation sickness and bone marrow suppression. The residents of two other islands, Ailinginae and Utirik, received significant whole-body doses as well—69 rem and 14 rem, respectively.

Of the 23 crewmembers of the *Lucky Dragon*, a Japanese fishing boat unexpectedly caught in the fallout from the Bravo test, one died and seven required intensive care for radiation sickness. The tuna on board were extremely contaminated. In 1954, Japanese monitoring programs showed that 683 fishing boats had contaminated fish in their holds, with 457 tons above acceptable levels.[35]

The Former Soviet Union

Between 1949 and 1989 the Soviet Union conducted 713 nuclear tests, most of them at Semipalatinsk in Kazakhstan and two sites on the Arctic islands of Novaya Zemlya. Nuclear tests were conducted at more than 50 other sites as well. The downwind communities in Kazakhstan were heavily contaminated. However, the studies conducted on the populations differ by wide margins in estimated exposure levels. A review of these differing studies suggests that over 10,000 people were exposed to

COHORT

CANCER SITE	Hanford	United Nuclear	Thorium Workers	Rocky Flats	Oak Ridge UCC
Oropharynx					
Esophagus					
Colon	●				
Rectum			●		
Liver/biliary			●	●	
Pancreas			○		
Lung	●		●		○
Bone		●			
Skin		●	●		
Prostate					
Kidney					
Eye					
Brain/CNS	●	●	○	●	○
Thyroid				●	
All lymphopoietic					
Lymphosarcoma			●		
Hodgkins			●		
Leukemia		●	●		
Other lymphatic	●				

● = SMR>100.   ○ = SMR>100 and statistically significant.

Figure 3   Index studies comparison of nuclear workers: elevated cancer sites by standardized mortality ratio (SMR). Adapted from references 21 and 41.

very high doses of radiation. Yet, medical data are insufficient to describe accurately the resulting radiation effects of these populations.[35]

Environmental contamination from nuclear weapons production by the former Soviet Union may even dwarf that of the United States. At Chelyabinsk (the official name was Mayak)—the USSR's main nuclear weapons production facility, radioactive wastes were routinely poured directly into the Techa River, exposing 124,000 people to increased levels of radiation. At the same site, a 1957 explosion that ripped through storage tanks containing radioactive wastes distributed 20 million curies of radionuclides over the surrounding countryside. Thomas Cochran, a senior scientist at the National Resources Defense

Pantex	Savannah River	Oak Ridge WWII	ORNL	Mound
		●		●
				●
	●		●	
		○		
		●		
				●
		●	●	●
				●
		●		
●		●	●	
	●			●
●			●	●
				●
●	●	●	○	●
	●			●

● = SMR>100.   ○ = SMR>100 and statistically significant.

Council in Washington, D.C., has called Chelyabinsk "the most polluted spot on Earth."[36]

Australia

Great Britain conducted 12 atmospheric nuclear tests at three locations in Australia between 1952 and 1957. This resulted in the dispersion of 1,350 curies of Pu-239, 32,000 curies of Cs-137, and 20,000 curies of Sr-90 across Australia. One-third of these amounts still remains in the environment.[35]

Algeria

France conducted 14 tests in the Sahara Desert between 1960 and 1965. Four were atmospheric and ten were underground. The French have not released any information that would allow a health assessment of the surrounding and downwind populations.[35]

China

China has conducted 22 atmospheric tests and 12 underground tests at Lop Nor in Sinkiang Province. The government admits to some deaths as a result of the testing, but no further information has been forthcoming. It is estimated that 3,300 curies of Pu-239 were dispersed into the atmosphere and 1,800 curies were deposited underground. Two million curies of Cs-137 and 1.3 million curies of Sr-90 were dispersed in the atmosphere.[35]

India

One underground nuclear test was conducted in the Thar Desert in 1974. No details are available. A suspected nuclear test in the South Atlantic near the Cape of Good Hope was detected by satellite in 1979. It has been surmised that this was a collaborative effort between South Africa and Israel.[35]

The Institute for Energy and Environmental Research has calculated the health effects of all atmospheric nuclear testing from the late 1940s to the year 2000, using data from the United Nations Scientific Committee on the Effects of Atomic Radiation (UNSCEAR)[37] and from the BEIR V report of the Committee on the Biological Effects of Ionizing Radiation. It has been estimated that there will be 430,000 excess cancer fatalities.[35] If these calculations are extrapolated to the entire period of time that global fallout from atmospheric testing will remain, a figure of 2.4 million excess cancer fatalities is obtained. Cancer incidence figures are many times higher than cancer fatality figures.

Difficulties in Assessing the Health Effects of Nuclear Weapons Production and Testing

Ever since the production and testing of nuclear weapons began, governments have deliberately withheld scientific and human health information from public view. Data regarding populations at risk, operational and accidental releases of hazardous chemicals and radionuclides, and waste disposal practices were classified. This secrecy has seriously hampered efforts at well-designed, reliable, and reproducible epidemiological studies. It has also deprived scientists of the knowledge necessary to understand the effects of low-level radiation on human health.[23] To date, most studies of nuclear workers or adjacent populations have been poorly designed or undermined by a lack of data.[21]

The science of risk assessment is not well developed. Owing to inadequate, unavailable, or lost exposure data, risk assessments typically rely upon dose reconstructions and modeling. In addition, risk estimates for radionuclides or hazardous chemicals are often based on best guesses or on minimal data on exposure (usually of animals). This practice often results in conclusions that fail to find an effect. The inability of risk-assessment studies to demonstrate a risk, however, does not mean that there is no risk. It only means that a risk may be present, but cannot be detected with current methodology.[22]

There is a growing body of evidence suggesting that there is no threshold level below which radiation exposure is safe.[10,21,38] Recent studies have suggested that chronic, low-dose radiation exposure has carcinogenic effects that do not become evident until years after the expected latency periods have elapsed. Thus, leukemias caused by these low-dose exposures may be seen 20 years, rather than the expected 10 years, later. Similarly, solid tumors may be seen 30 years, rather than 20 years, after exposure.[39]

Finally, the behavior, the transport, and the pathophysiology of radionuclides in human tissues are poorly understood. Radionuclides such as Pu-239, Cs-137, and Sr-90 are still being studied, and new hypotheses regarding their toxic effects are being proposed. Add to their carcinogenicity the effects of teratogenicity and chromosome effects on future generations, and it is apparent that we will continue to see the health effects of the production and testing of nuclear weapons for many generations to come.

Health Implications of Nuclear Power

Medical concern about nuclear power plants centers on four distinct aspects of their operation that endanger public health:

- low-level radiation exposures during routine functioning of the plants,
- the possibility of catastrophic reactor accidents,
- the problem of waste disposal and isolation, and
- the role of nuclear power in the proliferation of nuclear weapons.

Routine Operation

The major hazards to human health posed by some stages of the nuclear fuel cycle have been known for some time. As early as 1964, Wagoner et al. demonstrated a tenfold increase in fatal lung cancers among uranium miners.[9] Although the total number of persons involved in uranium mining is small, the extremely high risk of lung cancer among members of this profession raises serious ethical questions about allowing any individual to work as a uranium miner.

The extent of the dangers associated with other aspects of the nuclear fuel cycle—milling uranium ore, fabricating fuel rods, transporting fuel rod assemblies, and other operations—is less clear.[3,10,14,21,35,36] However, the data that are available, most of which come from workers in the U.S. nuclear weapons production program who perform tasks and receive radiation and chemical exposures similar to those of nuclear power plant workers, indicate that there may be significant health effects from these activities.

For example, a review of 10 cohort mortality studies of U.S. nuclear weapons workers exposed to cumulative radiation doses of less than 5 rem over their entire worklives (the current International Commission on Radiological Protection guidelines allow occupational exposure of 5 rem per year[40]) showed a rate of brain cancer 15% above that of unexposed workers.[41] A similar review of data from seven sites in the United States and one in the United Kingdom concluded that "considerable evidence now exists that nuclear workers do experience elevated risks from low dose external radiation for leukemia, lymphatic, and hematopoietic cancer combined, and to a lesser extent for solid tumors and for all cancers combined."[42] The risk, according to this review, may be 3 times that of unexposed workers for leukemia, 2.5 times for the lymph and blood cancers, and 1.7 times for all cancers and for solid tumors.

Aside from the danger to nuclear workers, there has been great public concern that low-level radiation emitted during the normal operation of nuclear power generating facilities might cause increased cancer rates and deaths in the general population. This concern has been particularly prevalent among populations living near commercial nuclear power plants. In the United States these fears have generally been dismissed on the grounds that plant emissions, as measured by physical dosimeters, do not exceed U.S. Nuclear Regulatory Commission (NRC) standards. There is considerable controversy, however, about the adequacy of these standards in protecting public health.[43,44]

The recent study of leukemia incidence around the Pilgrim nuclear power station in Plymouth, Massachusetts, has raised serious questions about the safety of reactors during routine operations. the study demonstrated a significant increase over expected levels in adult leukemias (other than chronic lymphocytic leukemia) among people living near the plant during times when released radiation levels did not exceed NRC limits.[43,44] Other studies on nuclear power plant emissions have demonstrated similar findings[45-47]; still others have found no increased cancer risk.[48,49] The impact of routine emissions from nuclear power plants clearly needs to be studied further to determine the precise nature of the health risks.

Catastrophic Accidents

The possibility of a reactor core meltdown accompanied by a catastrophic release of radiation has generated intense concern among the general public and members of the medical community. Since the beginning of the nuclear era, there have been numerous attempts to assess the magnitude of this danger.[50-53] Although these studies have all projected fatalities numbering in the tens of thousands and billions of dollars in property damage, they have also tended to minimize the risk of an accident's happening.

Two recent developments force a reevaluation of these early assessments of reactor safety.

The first is a growing awareness of the effects of aging on nuclear reactors. In addition to the usual deterioration that all industrial facilities experience as a result of heat, corrosive chemicals, and mechanical stress, nuclear reactors are subjected to the degrading effects of intense neutron irradiation, which makes the steel brittle and more likely to fracture.

This is particularly true for older reactors, whose nickel and copper alloy steels are more susceptible to embrittlement from neutron irradiation than are the purer steels in newer reactors. The prospect of aging nuclear reactors in their final years of operation is disturbing enough. But in the United States, as the oldest commercial reactors approach the end of their originally projected 40-year life spans, their operators have begun to apply for 20-year license extensions. It appears that these older reactors will not have to meet many of the safety standards that have been adopted since they were originally constructed.

The second is the 1986 accident at the Chernobyl nuclear power plant, which demonstrated conclusively that catastrophic reactor accidents were not merely theoretical possibilities. Millions of curies of radionuclides—including I-131, Sr-90, Ce-137, and Pu-239—were released into the surrounding countryside, contaminating 5.5 million hectares and exposing 2.5 million people to elevated radiation.[54] Hundreds of workers and others in the immediate vicinity received lethal doses of radiation, and many died from acute radiation sickness.[12,16,17] In 1990, Mettler et al.[55] found no increase in thyroid nodularity in residents of highly contaminated communities near the Chernobyl nuclear power station, compared to unexposed populations. However, a radiation scientist with the World Health Organization confirmed a dramatic increase in the incidence of thyroid cancer in children in the highly contaminated Gomel region of Belarus. In 1991 and the early months of 1992 the thyroid cancer rate in that region with 2.5 million inhabitants was 80 per million children—40–80 times the normal rate.[56] The Chernobyl accident deposited significant radioactive fallout over portions of Eastern and Central Europe, and in areas as far away as Sweden, Italy, and Wales. Agricultural and dairy products in many localities had to be destroyed.[4,16]

In the aftermath of Chernobyl, Commissioner James Asselstine of the U.S. Nuclear Regulatory Commission offered the following assessment: "Given the present levels of safety being achieved by the operating nuclear power plants in this country, we can expect to see a core meltdown accident within the next 20 years, and it is possible that such an accident could result in off-site releases of radioactivity that are as large as, or larger than, the releases estimated to have occurred at Chernobyl."[57]

Nuclear Waste Containment

A special health hazard associated with nuclear power is the production of vast inventories of radioactive wastes. These wastes consist of uranium mill tailings, transuranic wastes (defined as materials containing more than 100 nanocuries of transuranic material per gram of waste), "high-level wastes" (which include spent fuel rods and the sludges left behind when spent fuel rods are reprocessed to extract plutonium), and "low-level wastes" (which include those wastes not included in the other categories). "Low-level wastes" may include a wide range of materials, many of which are intensely radioactive. A typical 55-gallon drum of medical "low-level" waste will contain on the order of 10 millicuries. A typical drum of waste water from a nuclear reactor may contain 10 curies, and a drum of irradiated reactor components may contain 200–1,000 curies.

Commercial power reactors produce the vast majority of all "low-level" radioactive wastes, whether measured by volume or by total curies. In the area of "high-level wastes," the commercial nuclear power industry in the United States alone will have produced 41,000 metric tons by the year 2000.[57] There is no established safe technology for dealing with these wastes.

Low-level wastes have been stored at a number of waste dumps. The sites that have been studied most extensively are those associated with the nuclear weapons program. All of these have experienced major breaches of containment, with significant (in some cases, massive) contamination of surrounding areas. Current federal law requires states, acting individually or in cooperation with other states, to build a new generation of "low-level" waste site dumps, but there is no reason to believe that these will contain their wastes better than existing dumps.

In the United States, spent fuel rods are currently stored in cooling pools at reactor sites pending creation of an acceptable repository for them. No reported leaks from these storage pools have been reported, but there have been massive leaks from storage facilities for high-level waste from the nuclear weapons program.[58]

It is proposed that fuel rods from commercial reactors ultimately be stored in deep geological repositories, where, it is hoped, the waste will be safely isolated from the environment. However, no functioning repositories now exist. The proposed repository sites have all been criticized for failing to meet the criteria for geological stability, and the projected opening date for these facilities keeps receding as new problems are encountered. "The Department of Energy currently assumes that the earliest possible time at which the high-level waste repository at Yucca Mountain [Nevada] could be available . . . is the year 2015."[23] There is no guarantee, however, that this proposed site will actually be ready or will still be felt to be a suitable repository.

One unique aspect of the danger associated with radioactive wastes derives not only from the extreme toxicity of these materials but also from the great longevity of many of their isotopes. Cesium-137 and strontium-90 have half-lives of 30 and 28 years, respectively, and must be isolated for 300–600 years. Plutonium-239 has a half-life of 24,400 years and must be isolated for 240,000–480,000 years. The United Nations has been in existence only since 1945, and it has failed to prevent widespread civil and international wars even during this brief period. In the last 150 years, Australia and Antarctica are the only continents that have been spared widespread, violent social and political upheaval. All of recorded human history spans only 6,000 years. And yet we continue to produce huge quantities of highly toxic radioactive materials, knowing that to safeguard these materials would require an intact and functioning social order over thousands of years—something that has no precedent in human history. Never has a civilization deliberately and knowingly bequeathed to its descendants a burden of this magnitude.

Weapons Proliferation

As great as the above-mentioned threats are to public health, they pale in comparison with the greatest danger posed by nuclear power: its central role in the proliferation—actual and potential—of nuclear weapons. The recent fears about Iraq's attempts to develop nuclear weapons have underlined the dangers of nuclear proliferation. But the critical role of nuclear power in weapons proliferation has not received adequate attention, even though this role has been obvious from the outset of the nuclear era. In 1946, the Acheson-Lilienthal report warned that "the development of atomic energy for peaceful purposes and the development of atomic energy for bombs are in much of their course inter-changeable and interdependent."[59]

Highly enriched uranium, one of the two fissionable materials that can be used to make nuclear weapons, is not used in commercial reactors but is used in "peaceful" research reactors. Iraq's nuclear threat derived from its possession of 100 pounds of highly enriched uranium—enough to make two bombs equal in yield to the bomb that destroyed Nagasaki, and twice as much as the International Atomic Energy Agency believed it possessed.[60] This material was supplied by France and the Soviet Union for Iraq's "peaceful" research reactor. The United States, however, has supplied most of the 9,000 pounds of weapons-grade uranium that currently fuels 100 research reactors in 35 countries around the world.[60,61] This is enough to make 180 Nagasaki-size bombs.

Plutonium, the other fuel for nuclear weapons, is used directly in some commercial power plants. The growth of the plutonium-reproc-essing industry will lead to the production of vast quantities of this substance. Over the next 30 years, Japan alone could produce for its nuclear power industry 400 metric tons of plutonium—twice the amount currently in the combined arsenals of all the nuclear weapons powers.[61] Such inventories of plutonium will create extraordinary op-portunities for diversion to non-nuclear states intent on weapons devel-opment and to terrorist and criminal organizations.

As the Acheson-Lilienthal Report predicted in 1946, the "peace-ful" nuclear power industry has provided the rationale for would-be nuclear powers to acquire both the weapons-grade fuel and the cadre of skilled scientific and technical personnel needed to make atomic bombs.[59] It is difficult to imagine how we can hope to prevent the runaway proliferation of nuclear weapons as long as the nuclear power industry continues to produce the raw materials for weapons.

References

1. Mettler, F., Jr., and Ricks, R. C. Historical aspects of radiation accidents. In *Medical Management of Radiation Accidents,* ed. F. A. Mettler, C. A. Kelsey, and R. C. Ricks. CRC Press, 1990.

2. Health and Environmental Consequences of a Chernobyl Nuclear Power Plant Accident. Report DOE/ER-0332, Department of Energy, 1987.

3. Gofman, J. W. *Radiation-Induced Cancer from Low-Dose Exposure: An Independent Analysis.* Committee for Nuclear Responsibility, 1990.

4. von Hippel, F., and Cochran, T. B. Estimating long term health effects. *Bulletin of the Atomic Scientists* 15 (1986), no. 7: 18–24.

5. Aspaugh, L. R., Catlin, R. J., and Goldman, M. The global impact of the Chernobyl reactor accident. *Science* 242 (1988): 1513–1519.

6. Ginzburg, H. M., and Reis, E. Consequences of the nuclear power plant accident at Chernobyl. *Public Health Reports* 106 (1991), no. 1: 32–40.

7. Barnaby, F. Chernobyl, the consequences in Europe. *Ambio* 15 (1986): no. 6: 332–334.

8. Erikson, K. Toxic reckoning: Business faces a new kind of fear. *Harvard Business Review,* January–February 1990: 118–126.

9. Wagoner, J. K., et al. Cancer mortality patterns among United States uranium miners and millers, 1950–1962. *Journal of the National Cancer Institute* 32 (1964): 787–801.

10. Committee on the Biological Effects of Ionizing Radiation. *BEIR V: Health Effects of Exposure to Low Levels of Ionizing Radiation.* National Academy Press, 1990.

11. Biological effects of low-dose radiation: A workshop. *Health Physics* 59 (1990), no. 1: 1–102.

12. Abrams, H. L. Chernobyl: The emerging story. How radiation victims suffer. *Bulletin of the Atomic Scientists* 42 (1986): 13–17.

13. Caufield, C. *Multiple Exposures: Chronicles of the Radiation Age.* Harper & Row, 1989

14. Mancuso, T. F., Stewart, A. M., and Kneale, G. Radiation exposures of Hanford workers dying from cancer and other causes. *Health Physics* 33 (1977): 369–385.

15. Stewart, A. M., and Kneale, G. W. An overview of the Hanford controversy. *Occupational Medicine—State of the Art Reviews* 6 (1991), no. 6: 641–663.

16. Davis, A. M. Health care after Chernobyl: Radiation, scarcity, and fear. *Physicians for Social Responsibility Quarterly* 2 (1992): 3–24.

17. Medvedev, Z. *The Legacy of Chernobyl.* Norton, 1990.

18. Olshansky, S. J., and William, G. A comprehensive epidemiologic data resource. *Physicians for Social Responsibility Quarterly* 1 (1991): 145–156.

19. Gardner, M. W., Snee, M. P., and Hall, A. J. Results of case-control study of leukemia and lymphoma among young people near Sellafield Nuclear Plant. *British Medical Journal* 300 (1990): 423–429.

20. Evans, J. N. Gardner Report—Leukemia and radiation. *Nature,* May 3, 1990.

21. Geiger, H. J., et al. *Dead Reckoning: The Department of Energy's Epidemiologic Studies—A Critical Review.* Physicians for Social Responsibility, 1992.

22. Wilkinson, G. S. Epidemiologic studies of nuclear and radiation workers: An overview of what is known about health risks posed by the nuclear industry. *Occupational Medicine: State of the Art Reviews* 6 (1991), no. 4: 715–723.

23. Office of Technology Assessment, Congress of the United States. *Complex Cleanup—The Environmental Legacy of Nuclear Weapons Production.* Government Printing Office, 1991.

24. Schneider, K. U.S. admits peril of 1940s emissions at A-bomb plant. *New York Times,* July 12, 1990.

25. Davis, S., et al. Hanford Thyroid Disease Study Protocol Draft. CDC contract 200-89-0716, Fred Hutchinson Cancer Research Center, Seattle, 1990.

26. United States Department of Energy, Radionuclide Migration in Groundwater at the NTS. Nevada Operations Office, 1987.

27. Stevens, W., et al. Leukemia in Utah and radioactive fallout from the Nevada Test Site. *Journal of the American Medical Association* 264 (1980), no. 5: 585–591.

28. Anspaugh, L. R., et al. Historical exposures of external gamma exposure and collective external gamma exposure from testing at the Nevada Test Site. II. Test series after Hardtack II, 1958, and summary. *Health Physics* 59 (1990), no. 5: 525–532.

29. Cobb, J. Personnal communication. See also: Colorado Department of Health. Tritium Exposure and Bioassay Results for Residents of Broomfield, Colorado, 1976.

30. Rocky Flats Toxicology Review and Dose Reconstruction Project. Chemicals and Radionuclides Used at Rocky Flats. Repository document TW-362, 1991.

31. Johnson, C. J. Cancer incidence in an area contaminated with radionuclides near a nuclear installation. *Ambio* 10 (1981), no. 4: 176–182.

32. Barrick, C. W. Past Accidental Releases of Radioactivity from the Rocky Flats Plant. Environmental Sciences Department, Department of Energy, 1991.

33. Crump, K. S., Ng, T., and Cuddihy, R. G. Cancer incidence patterns in the Denver metropolitan area in relation to the Rocky Flats plant. *American Journal of Epidemiology* 126 (1987): 127–135.

34. Wilkinson, G. S., et al. Mortality among plutonium and other radiation workers at a plutonium weapons facility. *American Journal of Epidemiology* 125 (1987): 231–250.

35. Robbins, A. A., Makhijani, A., and Yih, K. *Radioactive Heaven and Earth—The Health and Environmental Effects of Nuclear Weapons Testing In, On, and Above the Earth.* Apex, 1991.

36. Hertsgaard, M. From here to Chelyabinsk. *Mother Jones,* January–February 1992.

37. United Nations Scientific Committee on the Effects of Atomic Radiation. *Ionizing Radiation: Sources and Biological Effects.* United Nations, 1982.

38. Greenberg, M. The evolution of attitudes to the human hazard of ionizing radiation and to its investigators. *American Journal of Industrial Medicine* 20 (1991): 718.

39. Wing, S., et al. Mortality among workers at Oak Ridge National Laboratory. *Journal of the American Medical Association* 265 (1991): 1397–1402.

40. International Commission on Radiological Protection. Recommendations of the ICRP. *Annals of the ICRP,* publication 26. Pergamon, 1977.

41. Alexander, V. Brain tumor risk among United States nuclear workers. In *The Nuclear Energy Industry,* ed. G. S. Wilkinson. Hanley and Belfus, 1991.

42. Wilkinson, G. S. Epidemiological studies of nuclear and radiation workers: An overview of what is known about health risks posed by the nuclear industry. In *The Nuclear Energy Industry,* ed. G. S. Wilkinson. Hanley and Belfus, 1991.

43. Clapp, R. W., et al. Leukemia near Massachusetts nuclear power plant. *Lancet* 2 (1987): 1324–1325.

44. Morris, M., and Knorr, R. S. Eastern Massachusetts Health Study, 1978–1986. Massachusetts Department of Public Health, Division of Environmental Health Assessment, 1990.

45. Forman, D., et al. Cancer near nuclear installations. *Nature* 329 (1987): 499–505.

46. Cook-Mozaffari, P. J., et al. Geographical variation in mortality from leukemia and other cancers in England and Wales in relation to proximity to nuclear installations, 1969–78. *British Journal of Cancer* 59 (1989): 476–485.

47. Ewings, P. D., et al. Incidence of leukemia in young people in the vicinity of Hinkley Point nuclear power station, 1959–1986. *British Medical Journal* 299 (1989): 289–293.

48. Jablon, S., Hrubec, Z., and Boice, J. D. Cancer in populations living near nuclear facilities. *Journal of the American Medical Association* 265 (1991): 1403–1408.

49. Enstrom, J. E. Cancer mortality patterns around the San Onore nuclear power plant, 1960–1978. *American Journal of Public Health* 73 (1983): 83–92.

50. Atomic Energy Commission. Theoretical Possibilities and Consequences of Major Accidents in Large Nuclear Power Plans. Report Wash-740, 1957.

51. Atomic Energy Commission and Brookhaven National Laboratory. Wash-740 Update, 1964–65.

52. Rasmussen, N. Reactor Safety Study—An Assessment of Accident Risks in United States Commercial Power Plants. Nuclear Regulatory Commission report Wash-1400, NUREG 75-014, 1975.

53. Strip, D. R. Estimates of the Financial Consequences of Nuclear Power Reactor Accidents. Nuclear Regulatory Commission report NUREG-CR 2723, 1982.

54. Shcherbak, Y. Keynote address, Health and Environment Conference, United Nations, New York, April 3, 1992.

55. Mettler, F. D., Jr., et al. Thyroid nodules in the population living around Chernobyl. *Journal of the American Medical Association* 265 (1992), no. 5: 616–619.

56. Baverstock, K. Thyroid cancer after Chernobyl. *Nature* 359 (1992): 21–22.

57. Geiger, J. H. Author's reply to a letter to the editor. *Journal of the American Medical Association* 257 (1982), no. 2: 190–191.

58. Department of Energy, Oak Ridge National Laboratory. Spent Fuel and Radioactive Waste Inventories, Projections, and Characteristics. Report DOE/RW-0006, 1986.

59. Department of State, Committee on Atomic Energy. Report on the International Control of Atomic Energy, 1946.

60. Leventhal, P. L., and Holland, D. J. Politicians in the lab. *Washington Post,* June 23, 1991.

61. Leventhal, P. L. *Latent and Blatant Proliferation: Does the NPT Work Against Either?* Nuclear Control Institute

62. Kossenko, M. M., Degteva, M. O., and Petrushova, N. A. Estimate of the risk of leukemia to residents exposed to radiation as a result of a nuclear accident in the southern Urals. *Physicians for Social Responsibility Quarterly* 2 (1992), no. 4: 187–197.

63. Kahn, P. A grisly archive of key cancer data. *Science* 259 (1993): 448–451.

War and the Environment: Human Health Consequences of the Environmental Damage of War

Jennifer Leaning, M.D.

. . . when in feudal times the aim of a king was to bring his truculent barons to heel, the primitive artillery of that period was found invaluable to deprive them of their power of resistance—their castles. But had its destructive effect been such that, not only their castles, but their retainers, serfs, orchards and cattle within a radius of several miles would be obliterated, nothing would have been left to bring to heel—the means would have swallowed the end.[1]

The human cost of war has been amply detailed in accounts that trace the toll exacted by all forms of weapons or other direct military action against human beings. The effect of war on the environment, as a general topic, has not received equivalent attention, and what notice it has attracted has arisen only recently. The subject of this chapter, the impact of war-induced environmental damage on human health, can be seen as an aspect of the general topic of war and the environment, and from this perspective it appears to have followed a similar time course of mild, late-blooming interest. Yet if the notion of environment is expanded to mean the human environment—the social, economic, and physical structures that constitute the niche in which human beings live and thrive—then it is evident that classic discussions of war have touched upon some of these factors (such as disease, famine, and exposure) that constitute secondary assaults on human health induced by the environmental damage of war.

The reasons for this relative delay in placing an analysis of war in an environmental context are beyond the scope of this discussion, except insofar as three developments contributing to the current high level of concern about the environment and its impact on human health all have recent roots in the years since the end of World War II.

- The evolution in the destructiveness of conventional weapons and the development of nuclear, chemical, and biological weapons have produced concrete evidence and theoretical grounds for the argument that the environment is vulnerable in many ways to the consequences of human aggression and can, in fact, be so damaged that it may fail to sustain the lives of those who survive the immediate, direct effects of the weapons.
- Military forces around the world have incorporated this technology-based potential for environmental destruction into modern war-fighting strategies, employing techniques of deliberate environmental destruction to weaken the defense and attack capacities of the enemy.
- The industrial and technical enterprises required to produce these highly sophisticated conventional weapons and weapons of mass destruction have drained societies of human and material resources, caused widespread environmental degradation, and inflicted serious harm on an unknown number of people.

There are four specific activities in the preparation or conduct of war that can be seen as harming the environment in ways that are particularly pernicious to human health: the production and testing of nuclear weapons, aerial bombardment, the planting of land mines, and the defoliation or despoilment of land, air, or water.

Production and Testing of Nuclear Weapons

The development during World War II of vast military industries capable of producing nuclear weapons has had a tremendously negative impact on local, regional, and global environments.[2,3] (See the preceding chapter for a more complete discussion of the environmental and health effects of nuclear weapons production and testing.)

In addition to the nuclear weapons complex, there are thousands of military bases and installations throughout the United States, which include more than two-thirds of the sites classified by the U.S. Environmental Protection Agency as highly toxic and dangerous.[4] Technologies for mitigating the dangers of the toxic chemical contamination created by these military sites are not well developed; the environmental (as well as social) costs of continuing to live with these toxicities and/or trying to contain them will undoubtedly be very high.

Aerial Bombardment

Bombardment of the human environment displaces the survivors, resulting in clusters of refugee populations in situations conducive to the spread of disease, to malnutrition and starvation, and to marked psychological stress. In World War II, air power became for the first time a major determining technology of military campaigns. During the 6 years of the war, the combatants expended approximately 6–9 million tons of air munitions on targets defined in broad terms of military usefulness.[5] These harbors, ports, overland rail and road routes, and industrial sites were also located near or within areas of dense human habitation. Hundreds of thousands of people died as a direct effect of these bombings. The indirect consequences brought on by the destruction of these human environments were also devastating: It is estimated that by the end of World War II approximately 40–50 million people in Europe alone were considered refugees—victims of a war so sweeping that it left people not only without homes but without countries.[6,7] The cumulative impact of this vast social dislocation on the subsequent course of world history has not been described.

During the years of active U.S. engagement in Southeast Asia, it is estimated that massive U.S. bombardment of Vietnam, Laos, and Cambodia forced approximately 17 million people to become refugees.[8] Thirty years later, with far more precision and efficiency, the Allied forces in the Gulf War relied on highly accurate and powerful aerial bombardment to destroy the urban human environment in Baghdad and several other major Iraqi cities. In a matter of weeks, water works, transportation systems, communications networks, and electrical power grids were selectively demolished. In a relatively urbanized country such as Iraq, this targeted destruction trapped the majority of the civilian population in a downward spiral of confusion, helplessness, hunger, and disease.[9–11] Iraqi military casualties have been estimated at 100,000 dead and 300,000 wounded.[12] According to a controversial U.S. Census Bureau analysis, Iraqi deaths arising from the direct and indirect effects of the ruin of the major cities may have totaled 100,000.[13]

Land Mines

Land mines were strewn across Europe and Africa during World War II, and in Vietnam and Cambodia during the Vietnam War to dissuade enemy movement across the mined terrain. For civilian populations who try to live on this land or to travel on mined waters once the war is over, unexploded mines and other munitions cast a long shadow across space and time. In Europe, in North Africa, and throughout Southeast Asia, intense efforts (often resulting in loss of life) have been required to clear and deactivate mines and other buried remnant munitions from arable land and pasture, and even after decades of such work

Figure 1   A victim of a land mine at the Thailand-Cambodia border. Courtesy of Rae McGrath, Mines Advisory Group.

millions of hectares remain under interdiction.[14] In Vietnam, Laos, and Cambodia, civilians are still frequently maimed or killed walking across territory thought to be free of such threats.[8,15] In Afghanistan, thousands of tons of mines and unexploded munitions litter the plains and the mountains, endangering the lives of farmers and herdsmen, their families, and their livestock for decades and greatly complicating the country's path toward peace.

Despoliation, Defoliation, and Toxic Pollution

Isolated instances of deliberately damaging the environment in time of war to inflict severe harm on the enemy can be traced back to antiquity. The primary means of doing this were to use fire or water to destroy supplies, ruin farmland, or block access routes.[16] During World War II there were two well-known examples of such efforts: the detonation of the Huayuankow dike across the Yellow River in China in 1938, whereby the Kuomintang hoped to block the advance of the Japanese but succeeded primarily in flooding several million hectares of farm land and drowning several hundred thousand Chinese,[17] and the opening of key dikes in the Netherlands in 1944, whereby the Germans intentionally flooded with salt water approximately 200,000 hectares of agricultural land.[18] Less widely known is the extent to which the German occupation forces, retreating under Russian attack in October 1944, devastated the human ecosystem of the northernmost settlements in Norway, slaughtering all the domestic animals, burning all the buildings, laying waste to all bridges, roads, and fishing boats, destroying all communications and utilities, and strewing the terrain and the harbors with mines.[19] During the Korean War, in May 1953, the U.S. deliberately bombed five irrigation dams in North Korea, all critical to rice production, in the hope of forcing the North Koreans to an armistice agreement.[20]

The war in Vietnam and parts of Laos and Cambodia, however, marked the first time a nation had used deliberate and direct destruction of the environment as a central and sustained facet of its war-fighting strategy.[8] The destruction of rice paddies in Vietnam was defended as necessary to deny the enemy its source of food; the destruction of rural areas in general was explained as needed to obliterate the enemy's protective cover.[8]

In the years 1965–1971, the U.S. sprayed 3,640 square kilometers of South Vietnam's cropland with herbicides, using a total estimated amount of 55 million kilograms.[8] The consequences of this onslaught on the ecosystem of Southeast Asia are only just beginning to be discerned. Little is known about the effects of this toxic exposure on the human population in this region, since ongoing war, civil strife, and diplomatic isolation have prevented internal or international epidemiological evaluations. An indication of potential morbidity for the Vietnamese population comes from the claims of U.S. veterans exposed to Agent Orange.[21]

The Gulf War, waged during January and February 1991, is remarkable for the extent of the environmental damage wrought in such a brief time frame. Early in the war, Iraqi forces released approximately 10 million barrels of Kuwaiti oil into the Persian Gulf,[22] which was already gravely polluted after decades of accumulating pollution from previous oil spills (particularly those from the Iraq-Iran war of 1980–88), from industrial wastes, and from heavy freighter traffic. The oil spill threatened Saudi desalination plants on the western shoreline, killed thousands of sea birds, and caused potential, as yet unquantified damage to seagrass beds and to a range of aquatic and migratory birds.

The Iraqi forces also set fire to 732 Kuwaiti oil wells,[23] filling the air for miles with dense black smoke that covered every surface with a sooty residue. The burning wells—not capped until November 1991—poured approximately 50,000 tons of sulfur dioxide and 100,000 tons of soot into the atmosphere per day.[25] Scientific assessments of the short- and long-term effects of this smoke have been filled with controversy and incomplete data,[26–29] but some preliminary measurements raise significant concerns about the potential health consequences for those downwind.[30,31] It has also been postulated that the short-term fluctuations in temperature and winds caused by the lofting smoke increased the intensity of the typhoon in Bangladesh.[27]

The military rationale for the Iraqi degradations of the local and regional environment seemed slim in Allied assessments, since the oil slick did not significantly impede Allied attack plans and the smoke from

Figure 2   Toxic cloud from a burning oil well, Kuwait, 1992. Source: International Committee of the Red Cross.

the Kuwaiti oil fires had no effect on the air or the ground war. An aspect of "eco-terrorism" has been alleged, in that Iraq is said to have sought to cause generalized psychological harm by manipulating the world's attachment to the environment.[30] The Iraqis, in turn, have charged the Allies with deliberately destroying their urban infrastructure in order to cause death and injury to the civilian population.

Major Issues in Estimating the Human Health Consequences of War-Induced Environmental Damage

Four main issues must be addressed in future work on this topic.

- Insufficient information exists about the effects of war on natural ecosystems, both in the immediate aftermath of war and over the long term. Until these effects are analyzed, assessments of the human impacts resulting from this environmental damage will continue to be fragmentary. Examples of the environmental effects of war, now accessible as case studies throughout the world, are not being studied in systematic or comprehensive ways. In many cases (such as the long-term effects of multiple craters or of disruption of desert terrain) there is a lack of clarity about what questions to ask or what data to gather. Public health experts, demographers, environmental scientists, meteorologists, agricultural experts, anthropologists, and military weapons experts have to begin to develop integrated analytic models to address these issues.

- The escalating numbers of weapons and the diverse technologies of destruction and delivery now available to virtually any country that wishes to pay the price place the local, regional, and global environments in greater jeopardy than ever before. Chemical and biological weapons, cluster bombs, fuel-air explosives, and herbicides are capable of inflicting massive and lasting damage on natural ecosystems and thus threatening the survival of entire human populations. Any war between forces equipped with these modern systems carries the potential of creating as much environmental destruction as the Gulf War.

- Burdened by rapid population growth in many parts of the world, by unrestrained settlement, and by economic exploitation, the global ecosystem is increasingly threatened and stressed, even in the absence

Figure 3   Nagasaki after the atomic bombing. Source: U.S. Air Force.

of war.[31] In view of the impact of World War II, another massive war in the near future (even one waged only with conventional weapons) would push into oblivion many natural systems that are now surviving marginally. The world's population has doubled since 1940, from 2.5 billion to over 5 billion. The rain forests have been reduced by more than 55% as of 1989 and are disappearing at a rate of 1.8% per year.[32] Global species extinction is occurring at such a pace that in 50 years one-fourth or more of the world's plants and animals could have disappeared.[33] Pollution of coastal oceans and seas is causing entire aquatic life chains to crash.[34] Air and ground pollution from industrial activity poses a serious health threat to many urban areas in developed and developing societies.[35] Since the end of the last world war, in times of relative peace, societies have approached the limits of global sustainability. A large-scale war occurring now, at this stage of environmental jeopardy, could plunge the world into ecological catastrophe.

· Future wars carry the risk that one party or more might use nuclear weapons. The environmental and human impacts of the use of nuclear weapons have been more carefully explored, on a theoretical basis, than have the impacts of the actual use of conventional weapons. Most assessments assume large-scale and widespread nuclear explosions, based as they are on strategic escalation scenarios. Even "limited" nuclear war, as envisioned during the depths of the Cold War for Europe (171 nuclear weapons, averaging 200 kilotons each, in the hypothetical scenario), would entail sweeping destruction of the environment from the North Sea to the Mediterranean and from England to the Urals, the death of approximately 10 million people in Germany alone, and enduring radioactive contamination.[36]

Global nuclear war would substantially destroy many life forms, particularly ocean plankton (from increased ultraviolet B radiation), deciduous trees and conifers (because of radiation sensitivity), and large mammals, including humans (again because of radiation sensitivity).[37,38] Longer-term effects of the war would include obliteration of most global agricultural production (through loss of human cultivators, destruction of seed, absence of petroleum and production facilities for fertilizer and mechanized equipment, radiation effects on crops that are planted, and global climate change precipitated by possible nuclear winter effects) (table 1). Such potentially profound alterations of the environment, described on the basis of well-

Table 1   Damage to biota from a nuclear bomb exploded at the surface. Source: reference 18.

Type of damage	Bomb size:	Area suffering the given type of damage (hectares)		
		18 kt	0.91 Mt	9.1 Mt
Craterization by blast wave		1	12	57
Trees blown down by blast wave		362	9,040	52,500
Trees killed by nuclear radiation		148	12,800	63,800
All vegetation killed by nuclear radiation		43	2,830	12,100
Dry vegetation ignited by thermal radiation		749	21,300	117,000
Vertebrates killed by blast wave		24	332	1,540
Vertebrates killed by nuclear radiation		674	36,400	177,000
Vertebrates killed by thermal radiation		1,000	26,900	150,000

grounded speculation, would cause from 1 billion to 4 billion human deaths from famine and disease.[39] Such assessments have served to caution the world community and to bolster efforts at arms control and disarmament. It is important to keep high on the agenda of world leadership the enormity of risk these weapons still entail.

International Law of War and the Environment[40-42]

The intense concern about the environmental effects of the Gulf War has been conducive to proposals to strengthen international law with regard to protection of the environment during time of war. International law has attempted to limit the use of weapons that carry high risk of damaging the environment and to prohibit the direct manipulation and destruction of the environment as an act of war. Of particular relevance to these efforts are the several existing treaties and arms control agreements that forbid the specific use of certain classes of weapons or certain methodologies: the 1925 Geneva Protocol on Chemical Weapons, the 1972 Biological Weapons Convention, the 1977 Protocols to the Geneva Conventions, the 1977 Convention on the Production of Military or any other Hostile Use of Environmental Modification Technologies (the "En-Mod Convention"), and the 1981 Convention on Excessively Injurious Conventional Weapons.[43]

Although these agreements mark important advances in the international rules of war as applied to the environment, they contain many loopholes and inadequacies. The challenge to the international community is to act, in the present favorable climate, to rectify some of the recognized insufficiencies in the current status of the law, and, in so doing, to introduce some new strictures. Experts in the field of arms control in relation to the environment have proposed the following[44]:

- ruling out military action if environmental consequences are assessed to be severe
- holding responsible each party that has caused environmental damage during armed conflict
- forbidding destruction or damage of facilities that could release radioactive or poisonous substances to the environment

- classifying natural parks and reserves and other sites of special ecological importance as demilitarized zones
- banning the use of all weapons of mass destruction.

Other proposals would not only increase our understanding of the
environmental effects of war but would also raise public consciousness
of the dangers of any specific impending armed conflict and support
mitigating interventions. Among these proposals[25] are that the United
Nations establish an international environmental database listing and
quantifying the details of vulnerable ecosystems, set up an environmental crisis management system that would preposition response
equipment (such as oil-spill rigs) and could dispatch teams of experts to
gather data and employ tactics to reduce environmental damage during
and immediately after a war, and require consideration of the environmental impact of a potential conflict as part of international decision-
making during times of crisis.

Conclusion

Although societies have waged war for millennia, the effects of this
enterprise on the environment and the pervasive consequences for human health have received relatively little sustained scientific attention.
The surge in interest after the Gulf War may reflect the recent overall
expansion of public awareness that the world's ecosystems are fragile,
even in times of peace, and that the fate of human beings is inextricably
bound to the fate of the earth. Such recognition has arrived late, only
as the environmental and social consequences of world population
growth and industrial activity have become too bleak to ignore.

To limit the environmental effects of war requires action at the
international level. The actions to take must spring from the tension that
pervades all international law with regard to war: bind all parties to
agreements that severely limit damage to the environment in the event
of war, and create norms, procedures, and alternatives that continue to
constrain the use of war as an option. Because the means of waging war
are increasingly outstripping the means of controlling its effects once
launched, the task for the United Nations in the 21st century is not only
to strengthen the laws regulating outright military action but also to
make possible and powerful all means to settle conflict short of war.

References

1. Fuller, J. F. C. *The Conduct of War 1789–1961*. Methuen, 1979.

2. Renner, M. G. Assessing the military's war on the environment. In *State of the World 1991*, ed. L. Brown. Norton, 1991.

3. Finger, M. The military, the nation state, and the environment. *Ecologist* 21 (1991): 220–224.

4. Environmental Progress and Challenges: EPA's Update. Environmental Protection Agency report GAO/RCED-88-44, 1988.

5. Westing, A. H. Misspent energy: Munition expenditures past and future. *Bulletin of Peace Proposals* 16 (1986), no. 1: 9–10.

6. Laquer, W. *Europe since Hitler*. Penguin, 1984.

7. Proudfoot, M. J. *European Refugees 1939–1952: A Study in Forced Population Movement*. Northwestern University Press, 1956.

8. Westing, A. H. *Warfare in a Fragile World: Military Impact on the Human Environment*. Taylor & Francis, 1980.

9. Arkin, W. M., Durran, D., and Cherni, M. *On Impact: Modern Warfare and the Environment; A Case Study of the Gulf War*. Greenpeace, 1991.

10. *Iraq Situation Report for Save the Children, U.K.* Save the Children Fund, 1991.

11. Renner, M. G. Military victory, ecological defeat. *World Watch*, July/August 1991: 27–33.

12. Tyler, P. E. Iraq's war toll estimated by U.S. *New York Times*, June 5, 1991.

13. Andrews, E. L. Census Bureau to dismiss analyst who estimated Iraqi casualties. *New York Times*, March 7, 1992.

14. Westing, A. H., ed. *Explosive Remnants of War: Mitigating the Environmental Effects*. Taylor & Francis, 1985.

15. Stover, E. *Land Mines in Cambodia: The Coward's War*. Physicians for Human Rights and Asia Watch, 1991.

16. Jones, A. *The Art of War in the Western World*. Oxford University Press, 1987.

17. Westing, A. H., ed. *Environmental Hazards of War: Releasing Dangerous Forces in an Industrialized World*. Sage, 1990.

18. Westing, A. H. *Weapons of Mass Destruction and the Environment*. Taylor & Francis, 1977.

19. Lund, D. H. The revival of northern Norway. *Geographical Journal* 109 (1947): 185–197.

20. The attack on the irrigation dams in North Korea. *Air University Quarterly Review* 6 (1954): 40–61.

21. Harris, R., and Paxman, J. *A Higher Form of Killing: The Secret Story of Gas and Germ Warfare.* Chatto & Windus, 1982.

22. Introductory report of the Executive Director: Environmental Consequences of the Armed Conflict between Iraq and Kuwait. United Nations Environment Programme report UNEP/GC.16/4/Add.1, 1991.

23. Carothers, A. After Desert Storm: The deluge. *Greenpeace,* October–December 1991: 14–17.

24. *The Gulf, 1990–1991: From Crisis to Conflict.* International Committee of the Red Cross, 1991.

25. Toukan, A. Humanity at war: The environmental price. *Physicians for Social Responsibility Quarterly* 1 (1991): 214–220.

26. Horgan, J. Up in flames. *Scientific American* 264 (1991), May: 17–24.

27. Horgan, J. Burning questions. *Scientific American* 265 (1991), July: 17–24.

28. Wald, M. L. Experts worried by Kuwait fires. *New York Times,* August 14, 1991.

29. Stone, R. Kuwait quits smoking. *Science* 255 (1992): 1357.

30. Environmental Consequences of the Armed Conflict between Iraq and Kuwait: Introductory Report of the Executive Director. United Nations Environment Programme report UNEP/GC 16/4/Add. 1, 1991.

31. Third National Toxics Campaign Fund Report on Kuwait Air Quality: Laboratory Test Results. National Toxins Campaign, 1991.

32. Ehrlich, P. R., and Wilson, E. O. Biodiversity studies: Science and policy. *Science* 253 (1991): 758–762.

33. Wilson, E. O. Threats to biodiversity. *Scientific American* 261 (1989), September: 108–116.

34. Ryan, J. C. Life Support: Conserving Biological Diversity. Paper 108, Worldwatch Institute, 1992.

35. Report of the WHO Commission on Health and the Environment: Summary. World Health Organization, 1992.

36. Arkin, W., von Hippel, F., and Levi, B. G. The consequences of a "limited" nuclear war in East and West Germany. In *The Aftermath: The Human and Ecological Consequences of Nuclear War,* ed. J. Peterson. Pantheon, 1983.

37. Pittock, A. B., et al. *Environmental Consequences of Nuclear War,* volume 1: *Physical and Atmospheric Effects.* Wiley, 1986.

38. Harwell, M. A., and Hutchinson, T. C. *Environmental Consequences of Nuclear War,* volume 2: *Ecological and Agricultural Effects.* Wiley, 1986.

39. Harwell, M. A., and Harwell, C. C. Nuclear famine: The indirect effects of nuclear war. In *The Medical Consequences of Nuclear War,* ed. F. Solomon and R. Q. Marston. National Academy Press, 1986.

40. DeLupis, I. D. *The Law of War.* Cambridge University Press, 1987.

41. *The Geneva Conventions of August 12, 1949.* International Committee of the Red Cross, 1989.

42. *Protocols Additional to the Geneva Conventions of August 12, 1949.* International Committee of the Red Cross, 1977.

43. Bouvier, A. Protection of the natural environment in time of armed conflict. *International Review of the Red Cross* 285 (1991): 567–578.

44. Barnaby, F. The environmental impact of the Gulf War. *Ecologist* 21 (1991): 166–172.

Loss of Stratospheric Ozone and Health Effects of Increased Ultraviolet Radiation

Alexander Leaf, M.D.

Atmospheric scientists were stunned in 1985 when a British Meteorological Survey team reported that concentrations of ozone over Antarctica had dropped by more than 40% during the early spring in the years between 1977 and 1984 from 1960 baseline levels.[1] Although it has been understood that gases liberated from commercial processes could reduce the stratospheric ozone layer, which forms a shield protecting all living things from the damaging effects of the ultraviolet (UV) radiations from the sun, there was no suspicion that the effect would be so drastic or occur so soon. It was thought the effect would be distributed rather evenly throughout the stratosphere, and that perhaps in 100 years this might result in a worrisome thinning of the protective shield. The discovery of the ozone hole was the first definitive evidence that human activities are changing the global environment, and this totally unsuspected finding attests to the limits of our present understanding of global environmental changes. In the early spring of 1987 and 1989, other research teams found that the ozone over Antarctica had dropped to only 50% of the 1979 levels and had become larger in area than the Antarctic Continent itself—about twice the size of the United States.[2,3] In March 1988, a definite (but lesser) thinning of the ozone layer over the North Pole was also noted.[4,5] And satellite measurements during the first three months of 1993 of the ozone layer over the heavily populated mid latitudes of the Northern Hemisphere have demonstrated record low levels.[24]

In the very cold, rarefied atmosphere of the stratosphere, some 20–50 kilometers above the surface, a molecule of oxygen (O_2) is split by solar ultraviolet radiation into two reactive oxygen units:

$$O_2 \rightarrow O + O.$$

If one of these oxygen atoms encounters a molecule of O_2, they may combine to form an ozone molecule:

$$O + O_2 \rightarrow O_3.$$

It is this thin layer of ozone above us that absorbs incident ultraviolet radiation, protecting animal and plant life on earth from its toxic effects.

Ultraviolet radiation is the portion of the electromagnetic spectrum from 200 to 400 nanometers. This region is further divided arbitrarily into subregions termed UV-A, UV-B, and UV-C. The last includes wavelengths of 200–290 nm and is most destructive to life; fortunately, it is effectively blocked from reaching the earth's surface by the atmosphere. UV-B includes solar radiations from 290 to 320 nm; it is many times more effective in inducing erythema than UV-A, which is 320–400 nm in wavelength. DNA and aromatic amino acids absorb UV-C maximally, UV-B significantly, and UV-A minimally. The pathologic consequences of UV radiations seem chiefly attributable to their absorption by and disruption of DNA and proteins. The damaging effects of UV-B exposure are cumulative and are not dependent on the rate of exposure; a dose fractionated over several days can be as deleterious as the same does delivered all at once.

Chlorofluorocarbons (which are used in aerosols, refrigerants, and other industrial products) are remarkably inert and nonreactive. Indeed, it is because of these characteristics—specifically because they are nontoxic and nonflammable—that they were invented. But when they eventually rise into the stratosphere, they are decomposed by solar ultraviolet radiations into free chlorine atoms:

$$\text{chlorofluorocarbon} \rightarrow Cl,$$

$$Cl + O_3 \rightarrow ClO + O_2,$$

$$ClO + O \rightarrow Cl + O_2.$$

The chlorine atoms are recycled in these reactions, and are then free to attack other ozone molecules. A single chlorine atom, released by the action of UV radiation on chlorofluorocarbons, is capable of destroying catalytically tens of thousands of ozone molecules during its residence in the stratosphere.

The release of chlorofluorocarbons into the atmosphere has increased dramatically in the last 30 years. Tropospheric levels of two of the most widely used CFCs, CFC-11 and CFC-12, have tripled or more during this period. Public and governmental concerns about CFCs led to the Montreal Protocol of 1987, which in turn led to the Helsinki Meeting in late 1988, at which it was proposed that chlorofluorocarbons and other ozone-depleting chemicals, such as methyl chloroform, be completely phased out by the end of 1999. Some countries have proceeded to ban all production of chloroflurocarbons even before this date. However, because of their long atmospheric lifetimes, it will be at least 100 years before the effects on the ozone layer of the chlorofluorocarbons that have already been liberated into the atmosphere disappear, even after all further production and use of chlorofluorocarbons is stopped. The reduction in ozone concentration over Halley Bay in Antarctica and its relation to the increasing levels of two chlorofluorocarbons (CFC-11 and CFC-12) are shown in figure 1.

There are now reasons to expect that positive feedback will amplify the ozone-destroying action of stratospheric chlorine. The seasonal and regional occurrence of ozone depletion are dependent chiefly on the catalytic photochemical reactions on the surfaces of microcrystals of ice in the very frigid temperatures of the stratosphere above the polar regions, when the first solar radiations return to the poles after the long polar winters. This is why the nadir in ozone concentrations occur in October over Antarctica and in March over the Arctic.

The eruption of Mount Pinatubo in June 1991 spewed 15,000–30,000 tons of sulfur dioxide (SO_2) into the stratosphere.[7] This was converted in about a month to sulfuric acid (H_2SO_4), which in the stratosphere condenses into small particles called *aerosols* that are expected to remain in the atmosphere for 1–3 years. The atmospheric aerosol load was increased thereby some 10–100 times over that produced by biological and human sources. The aerosol particles, like the microcrystals of ice, provide surfaces in the lower stratosphere on which the catalytic photochemical reactions that destroy the ozone can occur. Figure 2 shows the results of computer modeling of stratospheric ozone reductions as they vary by season and by latitude.

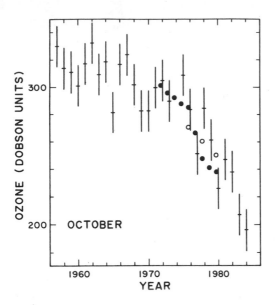

Figure 1   The average total ozone concentrations over Halley Bay, Antarctica (76°S. 27°E.) during the Octobers of 1957–1984. The circles represent observed concentrations of CFC-22 and CFC-12, graphed with increasing concentrations toward the bottom of the figure to indicate the correlation with ozone depletion. Source: reference 6.

Atmospheric scientists are monitoring these processes and measuring the increased incidence of UV-B reaching the earth's surface under the ozone holes. The World Meteorological Organization recently reported ozone levels over northern Europe, Russia, and Canada during the winter and spring of 1992 to have been 12% below the seasonal average—"an occurrence never before observed in more than 35 years of continuous ozone observations."[8] Thus, the depletion of the stratospheric protective ozone shield, first noted over Antarctica less than 10 years ago, has now spread over the northern hemisphere as well—in fact, over the entire globe. The depletion remains most severe over Antarctica, where the human population is small; however, the oceans there are rich in phytoplankton, which are the beginning of the food chain for all aquatic creatures. UV-B can penetrate several meters into the surface of the oceans, where the phytoplankton obtain the sunlight that is essential for photosynthesis. Phytoplankton are highly vulnerable to damage by UV-B,[9,10] so the potential for ecologic disaster is considerable.

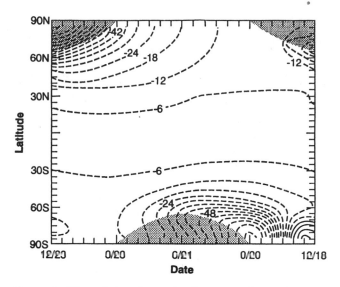

Figure 2   The relative change in the ozone density (in percent) at an altitude of 20 km, as a function of latitude and season during 1992. The reference case includes polar stratospheric clouds and background sulfate aerosols. Shaded areas represent polar night zones. Source: reference 7.

Health Effects of Increased Ultraviolet Radiation

The thinning of the stratospheric ozone shield and the resulting increase in UV-B radiation reaching the earth are expected to have direct health effects on humans. Exposure to the ultraviolet radiations in sunlight plays a major role in the premature appearance of aging of the skin. Dryness and wrinkles in the skin are brought on by solar exposure, and the preoccupation with acquiring a tan increases this cosmetic effect of sunlight.

UV-B and Skin Cancers

The incidence of all kinds of skin cancer, which constitute the commonest form of cancer among white populations, will increase with more exposure to ultraviolet-B.[11] Skin cancer has been divided into two forms: nonmelanoma skin cancer, which affects the principal cell type of the skin (the keratinocyte), and cutaneous melanoma, which affects the pigment-producing cell (the melanocyte). Nonmelanoma skin cancers are of two kinds: basal-cell and squamous-cell carcinoma. Melanoma is the most dangerous form of skin cancer; it is estimated that in any year about 25% of the Americans who develop melanoma die from it. Basal-cell and squamous-cell cancers are much less dangerous, with a combined mortality of less than 1% (most of it due to squamous-cell carcinoma).

The incidence of melanomas has already been increasing in the United States; between 1982 and 1989 it rose by 83%. Unlike basal-cell and squamous-cell carcinomas, which seem to increase in proportion to total exposure to ultraviolet radiation, melanoma appears to be associated with an acute exposure such as a severe sunburn, which may serve as the initiating stress. Years or decades later, some other stress may serve to start the melanoma. Whether the second stress is also exposure to solar radiation is not known. The extent to which public education will offset this increase, by encouraging the wearing of hats and other protective clothing and the use of sunscreen lotions, is unpredictable. The U.S. Environmental Protection Agency estimates that if ozone depletion is allowed to continue until there is 40% depletion (estimated to occur in 2075), the increase in biologically active UV would result in an additional 154 million cases of skin cancer and an additional 3.4 million deaths.[11,12] (This estimate was for individuals alive at the beginning of the assessment and born before 2075.)

Basal-cell cancers of the skin are very common today among fair-skinned individuals who imprudently expose themselves to excessive sunlight. These cancers do not metastasize to distant parts of the body or to internal organs, as do melanomas; they enlarge slowly and are cosmetically unsightly. Squamous-cell cancers are uncommon. They grow faster than basal-cell tumors and have the potential for metastasis. Individuals with dark skins are resistant to these solar-initiated cancers, and blacks are essentially immune. The dark melanin pigment in their chromophore cells within the skin absorbs the incident ultraviolet light, thus protecting the underlying skin cells from irradiation.

UV-B's Effects on the Eye

Cataracts may be a more widespread health effect of ultraviolet-B radiation than skin cancers, because all populations will be affected. A cataract occurs when the normally translucent lens of the eye becomes cloudy and scatters light so that vision is impaired. Cataracts account for half of the blindness in the world,[13] amounting to some 20 million cases. About 1.25 million cataract operations are performed annually in the United States, where cataracts currently represent the third-largest cause of preventable blindness.

Exactly how the ordered crystalline proteins of the lens become denatured, causing a cataract, is still uncertain. The lens proteins contain amino acids, such as tryptophane, which are susceptible to photo-oxidation by oxygen free radicals generated by UV-B. It is believed that the oxidized lens proteins become structurally altered.

Studies of fishermen and others engaged in gathering seafoods in Chesapeake Bay, who were exposed not only to direct solar radiation but also to its reflection from the ocean surface, showed a higher incidence of cataracts than did a control group who worked entirely indoors. Interestingly, wearing glasses or sunglasses is sufficient to block much of the ultraviolet radiation that causes cataracts. Plastic lenses, whether tinted or clear, reduce ocular exposure by more than 90%. Glass lenses afford only about 80% protection. Hats as well as protective eyewear provide major protection. Hugh R. Taylor and associates reported that people who worked indoors have a typical exposure of 4 units (Maryland Sun Years).[14] The typical waterman who worked outside without any ocular protection had an ocular exposure 18 times higher (72 units). If he wore a hat, his ocular exposure was cut almost in half, to 47 units. If he wore glasses, it was cut to 17 units. Both a hat and glasses reduced his ocular exposure to only 8 units, only twice that of someone who was spending essentially all day inside. Furthermore, logistic regression analysis controlling for age showed a strong correlation between cumulative UV-B exposure and cataract. This analysis indicated that a doubling of UV-B exposure will increase the risk of developing cataracts by 60%. Conversely, if the ocular exposure to UV-B is halved, the risk is reduced by 40%. Those in the top quartile of annual ocular exposure had more than 3 times the risk of developing cortical cataract of those in the bottom quartile. There was no indication of any safe exposure or threshold for UV-B exposure, or of any safe period for exposure.

Epidemiologic studies have indicated that a diet rich in natural anti-oxidants, including vitamins C and E and beta carotene, reduces the incidence of cataracts.[15] This would support the hypothesis that oxygen free radicals play a causative role in cataract formation. To a considerable extent, the loss of vision due to cataract is correctable today, with surgery and the implantation of prosthetic lenses. But such operations are available to only a small percentage of those affected, particularly in the developing world.

Photokeratitis, or snowblindness, is another adverse effect of UV-B on the eye. Photokeratitis—literally sunburn of the cornea—results largely from UV-B exposure.[16] After undue exposure to bright sunlight, the afflicted individual develops red, inflamed, irritable, sore eyes. Photokeratitis clearly occurs with current UV-B levels, especially over bright snowy surfaces. It seems that we are very close to the threshold level for photo-keratitis during the summer months, especially at the beach. It is not known what level of depletion of stratospheric ozone and accompanying increase in ambient UV-B can be tolerated before photokeratitis becomes a much more widespread and common problem.

Effects of UV-B on the Immune System

The effects of ultraviolet-B radiation on the skin, in addition to cancer formation, have been studied intensively in recent years. It is now recognized to have an important effect on the immune system.[17]

There are two major arms to the body's defense against foreign antigens: a humoral and a cellular defense. The humoral immune system produces antibodies (immunoglobulins) that can react very specifically with and neutralize the harmful effects of foreign molecules that gain access into our bodies. The humoral immune system appears, however, to be unaffected by ultraviolet radiation. By contrast, the cellular immune system, which recognizes a cancer cell or a parasitic cell and mounts an immune attack, ultimately resulting in the killing and phagocytosis of the offensive foreign invader, is sensitive to ultraviolet radiation. The central player in the cellular immune system is the T-lymphocyte. Exposure to sunlight alters subsets of T-cells and induces suppressor T-cell activity in normal subjects.[18]

As described above, ultraviolet radiation can cause the development of malignant skin tumors. In experimental animals it has been shown that cancer formation requires uniformly large doses of irradiation.

When such tumors, induced by ultraviolet irradiation, are transplanted into normal genetically identical recipient animals, however, they are unable to grow. If the recipient animal is first exposed to a much smaller dose of ultraviolet radiation than that required to induce a skin tumor, the transplanted tumor is then not rejected. This clearly indicates that somehow the small exposure of the recipient animal to ultraviolet radiation has depressed the animal's immune system so that it will tolerate the growth of the cancer implant. Furthermore, this depression of the recipient animal's immune system must be a systemic effect, since the cancer implant will grow in the recipient even if transplanted to a site distant from that which was exposed to the ultraviolet radiation. Ultraviolet radiations are known to penetrate only the outermost layer of the skin.

All the details of how exposure to ultraviolet radiation may suppress the cellular immune responses are not yet elucidated. It is known that there are several cells of the immune system that are transiently resident in the skin but arise from tissues elsewhere. These include, beside the lymphocytes, macrophages and Langerhans cells that originate in the bone marrow. Both the Langerhans cells and the dermal macrophages are antigen-presenting cells, but they also secrete several chemical messengers that regulate other immune cells. Exposing normal skin to low doses of ultraviolet radiations results in a loss of recognizable Langerhans cells at the site of the ultraviolet exposure. Thus local suppression of cellular immunity can be explained.

The systemic suppression of the immune system appears to result chiefly from the altered production of chemical messengers arising from the effects of UV-B on the several important cells of the immune system that reside in the skin. These potent messenger molecules modulate the responses of the immune system throughout the body, and in consequence of exposure to UV-B they suppress the cellular immune responses.[19]

Under conditions expected to prevail with global environmental changes, this suppression of the immune system acquires special significance. Since UV-B exposure can initiate skin cancers as described, a deficient cellular immune response to the presence of the cancer cells can only result in their more rapid proliferation and spread. More importantly, however, will be the impaired response of exposed individuals to infections, increasing the incidence of infections and making them more lethal. With the crowding due to population increases and to the displacement of people from submerged coastal and drought-stricken areas (secondary to global warming), with the mixing of immune and non-immune populations, with the possible spread of vector-borne diseases into new territories, with the contamination of water supplies, and with the increased prevalence of malnutrition and poverty, most kinds of infectious diseases will likely be increased. For people to have their immune systems suppressed under such conditions can only greatly aggravate the serious problems that will be posed by infectious diseases.

Effects of Tropospheric Ozone on Respiratory Function

Closer to the ground, ozone, a major component of smog, is produced through chemical reactions involving hydrocarbons and nitrogen oxide emissions from automobiles and industrial processes. With reduction in the stratospheric ozone shield, more UV-B will penetrate down into the troposphere, where it will produce ozone from molecular oxygen just as it does in the stratosphere. It has been shown in laboratory animals and in humans that ozone is a respiratory irritant[20] that affects lung function in a manner similar to cigarette smoke, causing constriction of the small bronchial airways and a reduction in lung capacity. This causes increased predilection to respiratory infections, emphysema, and chronic pulmonary insufficiency.

Ozone plays two contrasting roles in the expected global environmental changes. In the stratosphere it provides a shield against harmful UV-B radiations, while in the troposphere it acts as a greenhouse gas and as a health hazard. So we are losing the ozone where it is beneficial and gaining it where it is deleterious to human health.[21]

Effects of Increased UV-B on Food Production

Of all the health effects of the anticipated global environmental changes, food shortages are likely to be the most devastating. UV-B radiations are toxic to most plants[22] and will adversely affect agricultural productivity. As mentioned, phytoplankton, which are abundant in the cold polar oceans, are impaired by ultraviolet radiation. Since the phytoplankton form the beginning of the food chain on which all marine species depend, the possibility of ecologic catastrophe looms. Tropospheric ozone, an indirect result of depletion of stratospheric ozone, is also toxic to most plants. Levels of ozone over the eastern United States during summers are already high enough to cause damage to crops and vegetation. These effects of UV-B on agriculture will only add to the many other factors that will compromise food supplies for the world's burgeoning population, especially in the developing countries which already are faced with marginal supplies.[23]

References

1. Farman, J. C., Gardiner, B. J., and Shanklin, J. Large losses of total ozone in Antarctica reveal seasonal ClO_x/NO_x interaction. *Nature* 315 (1985): 207–210.

2. Shell, E. R. Solo flights into the ozone reveals its causes. *Smithsonian,* February 1988: 142–155.

3. Kerr, R. A. Stratospheric ozone is decreasing. *Science* 239 (1988): 1489–1491.

4. Solomon, S., et al. Observations of the nightime abundance of OCLO in the winter stratosphere above Thule, Greenland. *Science* 242 (1988): 550–555.

5. Mount, G., et al. Observations of stratospheric NO_2 and O_3 at Thule, Greenland. *Science* 242 (1988): 555–558.

6. Rowland, F. S. Stratospheric ozone depletion. *Annual Review of Physical Chemistry* 42 (1991): 731–768.

7. Brasseur, G., and Granier, C. Mount Pinatubo aerosols, chlorofluorocarbons, and ozone depletion. *Science* 257 (1992): 1239–1242.

8. United Nations agency cites lowest recorded levels for ozone. *Boston Globe,* November 14, 1992.

9. Roberts, L. Does the ozone hole threaten Antarctic life? *Science* 244 (1989): 288–289.

10. Bridigare, R. R. Potential effects of UVB on marine organisms of the southern ocean: Distribution of phytoplankton and krill during austral spring. *Photochemistry and Photobiology* 50 (1989): 469–478.

11. Longstreth, J. Skin cancer and ultraviolet light: Risk estimates due to ozone depletion. In *Global Atmospheric Change and Public Health,* ed. J. C. White. Elsevier, 1990.

12. Brown, L. R., ed. *State of the World 1989.* Norton, 1989.

13. Taylor, H. R., et al. Effects of the ultraviolet radiation on cataract formation. *New England Journal of Medicine* 319 (1988): 1429–1433.

14. Rosenthal, F. S., et al. The effects of sunglasses on ocular exposure to ultraviolet radiation. *American Journal of Public Health* 78 (1988): 72–74.

15. Jacques, P. F., and Chylack, L. T., Jr. Epidemiologic evidence of a role for the antioxidant vitamins and carotenoids in cataract prevention. *American Journal of Clinical Nutrition* 53 (1991), suppl. 1: 352S–355S.

16. Taylor, H. R. Cataracts and ultraviolet light. In *Global Atmospheric Change and Public Health,* ed. J. C. White. Elsevier, 1990.

17. Gallo, R. L., Staszewski, R., and Granstein, R. D. Physiology and pathology of skin photoimmunology. In *Skin and Immune System,* ed. J. Bos. CRC Press, 1990.

18. Hersey, P., et al. Alteration in T cell subsets and induction of suppressor T cells activity in normal subjects after exposure to sunlight. *Journal of Immunology* 31 (1983): 171–174.

19. Daynes, R. A. Immune system and ultraviolet light. In *Global Atmospheric Change and Public Health,* ed. J. C. White. Elsevier, 1990.

20. Whittemore, A. S. Air pollution and respiratory disease. *Annual Review of Public Health* 2 (1981): 397–429.

21. Schneider, T., ed. *Atmospheric Ozone Research and Its Policy Implications.* Elsevier, 1989.

22. Worrest, R. C., and Caldwell, M. M., eds. *Stratospheric Ozone Reduction, Solar Ultraviolet Radiation and Plant Life.* Springer-Verlag, 1986.

23. *Report of the Task Force on Food Security in Africa.* World Bank, 1988.

24. Northern hemisphere ozone at 14-year low. *New York Times,* April 23, 1993.

The Possible Effects of Climate Change on Health

Andrew Haines, M.D.

A number of predictions have been made about the increase in temperature that is likely to occur over the next century or so as a result of the steady accumulation of greenhouse gases in the atmosphere. The scientists on the Intergovernmental Panel on Climate Change (IPCC) examined a number of scenarios and calculated that, with "business as usual," the global mean temperature will increase by about 0.3°C per decade (with a range of 0.2–0.5°C per decade). This will lead to an increase in the global mean temperature of about 1°C above the present value by 2025 and 3°C (with an upper estimate of 5°C) by 2100. These are realized rises in temperature, and they account for only about 60–80% of the final rise because the thermal inertia of the oceans acts as a buffer.[1] Figure 1 compares the projected speed of climate change according to the IPCC scientists with past climatic events.

A number of feedback mechanisms can come into play, and these can either increase or reduce change. The majority seem likely to accentuate the warming process. One example is the potential release of methane from permafrost or from under the oceans with rising temperatures.[1] There is concern that the predominance of feedback mechanisms tending to accelerate climate change would result in more rapid warming than the IPCC predictions suggest. Land surfaces warm more rapidly than oceans, and the higher latitudes, particularly in winter, warm more rapidly than the tropics. Confidence in the prediction

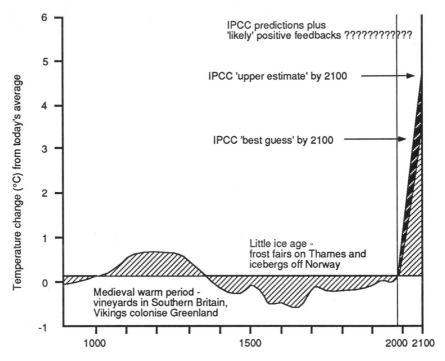

Figure 1 Change in global temperature up to 2100. Adapted, with permission, from J. Leggett, "The nature of the greenhouse threat," in *Global Warming: The Greenpeace Report,* ed. J. Leggett (Oxford University Press, 1990).

of regional changes is low, but it seems likely that there may be reduced summer rainfall and soil moisture in southern Europe and central North America. Under the IPCC's "business as usual" scenario, global mean sea level is predicted to rise by an average of about 6 centimeters per decade over the next century (with a range of 3–10 cm per decade), mainly because of thermal expansion of oceans and melting of land ice.

Global mean surface air temperature has increased by 0.3–0.6°C over the last 100 years, but the unequivocal detection of an enhanced greenhouse effect due to emissions related to human activities is not likely for a decade or more. Around 50% of the greenhouse effect is caused by carbon dioxide; chlorofluorocarbons, methane, and nitrous oxide account for most of the remainder. A summary of major green-

Table 1 Key greenhouse gases influenced by human activities. Adapted from reference 1 with permission from Cambridge University Press.

Parameter	Carbon dioxide	Methane	CFC-11	CFC-12	Nitrous oxide
Pre-industrial atmospheric concentration (1750–1800)	280 ppmv[a]	0.8 ppmv	0	0	288 ppbv[a]
Atmospheric concentration (1990)[b]	353 ppmv	1.72 ppmv	280 pptv[a] 484 pptv	310 ppbv	
Current rate of annual atmospheric accumulation	1.8 ppmv (0.5%)	0.015 ppmv (0.9%)	9.5 pptv (4%)	17 pptv (4%)	0.8 ppbv (0.25%)
Atmospheric lifetime[c] (years)	(50–200)	10	65	130	150

Ozone has not been included in the table because of lack of precise data.

a. ppmv = parts per million by volume; ppbv = parts per billion by volume; pptv = parts per trillion by volume.

The 1990 concentrations have been estimated on the basis of an extrapolation of measurements reported for earlier years, assuming that the recent trends remained approximately constant.

For each gas in the table, except CO_2, the lifetime is defined as the ratio of the atmospheric content of the total rate of removal. CO_2 is a special case since it has no real sinks, but is merely circulated between various reservoirs (atmosphere, ocean, biota). The "lifetime" of CO_2 given in the table is a rough indication of the time it would take for the CO_2 concentration to adjust to changes in the emissions.

house gases is shown in table 1. As figure 2 indicates, the atmosphere is gaining around 3 gigatonnes (Gt) of carbon per year.

The IPCC's 1992 update of its 1990 report suggests that the contribution of CFCs to the greenhouse effect may be at least partially offset by the depletion of ozone in the lower stratosphere at middle and high latitudes.[3] In addition, the cooling effect of aerosols from sulfur emissions from the burning of fossil fuels and from the June 1991 eruption of Mount Pinatubo in the Philippines may have offset part of the greenhouse warming in the northern hemisphere. In general, though, the IPCC confirmed its earlier conclusions.

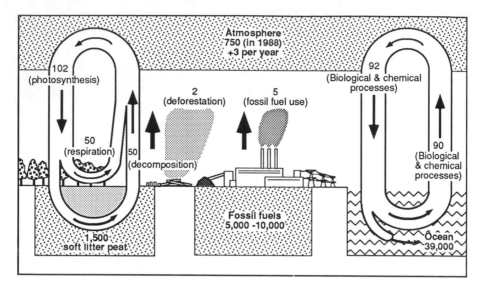

Figure 2 Generalized portrayal of the carbon cycle, showing main reservoirs and fluxes. Quantities of carbon are in Gt (reservoirs) and Gt per annum (fluxes). Source: J. Leggett, "The nature of the greenhouse threat," in *Global Warming: The Greenpeace Report,* ed. J. Leggett (Oxford University Press, 1990); adapted from S. Schneider, "The changing climate," *Scientific American* 261 (1989), no. 3: 70–79, with permission of Oxford University Press and Scientific American, Inc.

The potential health impacts of climate change could be both direct and indirect. Although much of the literature up to now has focused on the direct effects of temperature on health, the indirect effects will probably be more important overall. Even though it is impossible to make accurate predictions of potential impacts on human health, it is important to consider the possible consequences of climate change, in order to assess ways in which human populations may be particularly vulnerable so that countermeasures may be considered in advance and early signs that climate change is having an impact on human health may be detected. In addition, it is likely that the potential for adverse impacts on human health will have a strong influence on the opinion of the public and that of policy makers. The difficulty of predicting impacts on human health in quantitative terms at this stage is not an argument for inactivity. Rather, it suggests that policy makers should take a precautionary approach so as to limit greenhouse-gas emissions and minimize climate change.

Direct Effects of Temperature Increases

Increases in temperature may affect a number of major categories of disease, particularly cardiovascular, cerebrovascular, and respiratory diseases. For instance, a study of the relationship between temperature and vascular disease mortality in several U.S. cities demonstrated an inverse relationship between about −5°C and about +25°C,[4] the number of deaths tending to decline with increasing temperature. Above and below this range there were increases in mortality, particularly among the elderly. The increase in mortality from stroke above 25°C was particularly marked in many of the cities. The relationship between cardio- and cerebrovascular mortality and temperature is shown in figure 3 for 12 U.S. cities.[4]

In the United States, under conditions of doubled atmospheric carbon dioxide concentrations, the number of summer deaths due to high temperature has been predicted to rise from an estimated current total of around 1,150 to around 7,400. About 60% of the deaths are predicted to occur among persons aged 65 and over. It has been suggested that, even with acclimatization, the number of summer deaths associated with the rising temperature is likely to exceed greatly the reduction of winter deaths.[5] During three heat waves in Los Angeles when temperatures were about 41°C, the peak mortality at all ages was between 172% and 445% that expected.[6] Among persons aged 85 and over the peak mortality was even higher—between 257% and 810% that expected. Associations between increases in daily mortality and short spells of relatively hot weather have also been described in greater London.[7] Peak mortalities from coronary and cerebral thromboses occurred one or two days later.

A study in which small numbers of volunteers were exposed to moving air at 41°C for 6 hours caused rising core temperatures of 0.8°C, a fall in weight of 1.8 kilograms despite access to water, a rise in heart rate of around 30 beats per minute, and a fall in blood pressure (particularly on standing). Blood viscosity increased by 24%, and the platelet count rose 18%. Plasma cholesterol increased by 14%. The increases in whole-blood viscosity and plasma cholesterol were thought to be explained by a reduction in plasma volume. This was attributable

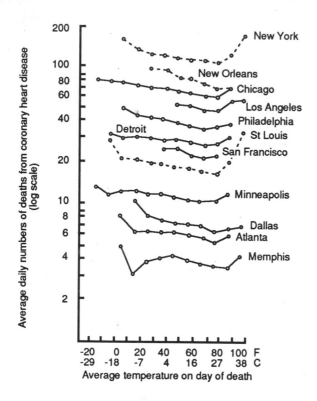

to an isotonic loss of salt and water from the body. However, the increase in platelet count was double the decrease in blood volume, indicating release of platelets into the circulation.[8]

Acclimatization to high temperatures may ameliorate the effect of temperature, but the elderly and those with cardiovascular and other major diseases may adapt less well. In urban areas, heat waves are unlikely to cause severe heat exhaustion, because it may be possible to restrict physical activity outdoors and to have ready access to fluids. However, where potable water supplies are limited and cooling systems are inadequate, the availability of clean water will become critical for human survival. In extreme situations, heat exhaustion syndromes and heat intolerance may lead to heat stroke, characterized by core temperatures of 41°C or more, with central-nervous-system disturbances causing convulsions and coma.

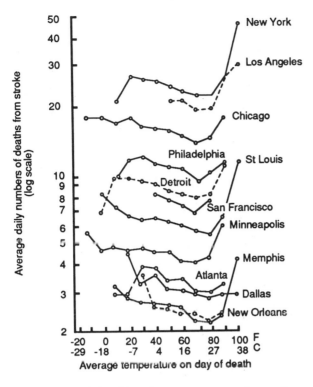

Figure 3 Relationship of temperature to heart disease and stroke mortality. Source: A. Haines, "The implications for health," in *Global Warming: The Green-peace Report*, ed. J. Leggett (Oxford University Press, 1990); reproduced with permission of Oxford University Press and *American Journal of Epidemiology*.

Behavioral symptoms due to prolonged high temperatures in urban environments, which can include social intolerance, irritability, and industrial accidents, may be more widespread (though less serious). Urban areas may have significantly higher temperatures than surrounding localities because of the retention of heat by buildings.[9] In general, elderly people—whose numbers are increasing in many developed countries—will suffer the most serious consequences of increased temperatures. Clearly, the overall impact will depend on the degree to which reductions in mortality due to cold weather are counterbalanced by increases due to hot weather, which in turn are likely to vary according to geographical location.

Indirect Effects of Climate Change

Communicable Diseases

There are several mechanisms by which the incidence of communicable diseases may be affected. There will be effects on vectors, which will influence the distribution of diseases that are currently prevalent mainly in tropical and subtropical regions. There may also be direct effects on the quality of water for drinking and cooking and changes in soil characteristics, which may in turn affect helminth (flatworm) infections. Population movements may facilitate the spread of communicable disease; for example, when large numbers of refugees are gathered together, sanitation is often poor and increased person-to-person transmission of disease may occur.

An increase in temperature may cause an expansion or a shift in the distribution of vectors to higher latitudes or altitudes. It could also cause acceleration of the development of parasite stages in a vector. For example, the extrinsic incubation period for the dengue virus and the yellow fever virus decreases with increasing mean ambient temperature.[10,21] In the case of malaria, the length of the development cycle for the parasite mosquitoes also varies with temperature, the shortest cycle occurring at 26°C.[11] These effects on incubation periods for the infectious agents, coupled with a wider distribution and a more rapid metamorphosis of the mosquitoes, all of which may result from increasing temperatures, could lead to faster-growing and more widespread epidemics.[21]

Large numbers of individuals are currently at risk from major vector-borne diseases. The population at risk from malaria is 2.1 billion, that from schistosomiasis 600 million, and that from lymphatic filariasis 100 million, the respective prevalences being 270 million, 200 million, and 90.2 million.[11] The distribution of all three of these conditions is likely to be affected by global warming. In areas which border malarial regions or in which malaria is currently endemic, climate change could lead to increased density of vectors. In Egypt, which lies partly on the edge of hot and temperate zones, water snails tend to lose schistosome infections during winter months. This tendency may be diminished or abolished by global warming. Chagas' disease is related to human activities and habitation; however, in areas where reproduction of vectors and transmission are seasonal, the reproduction rates will increase as temperatures rise.

Drought and desertification in some areas will result in a reduction in density of water-related vectors, but may increase the transmission in these regions of diseases such as dracunculiasis (guinea-worm) and perhaps African trypanosomiasis (if, for instance, people and animals are gathered together for increasing periods of time at limited numbers of watering holes). The WHO Task Group did not, however, consider that changes in the distribution of dracunculiasis were likely.[11] Changes in vegetation may affect the distribution of ticks and therefore the distribution of a range of tick-borne diseases. Coastal flooding might lead to the formation of more brackish lagoons, extending the breeding of some types of brackish-water species (such as *Anopheles sundaicus* in Indonesia).

Some affluent countries, such as Australia and the United States, may also suffer increased threats from communicable disease as a result of climate change. In Australia, there may well be an increase in the instance of epidemic polyarthritis (caused by the Ross River virus). An extension of the area in which the Murray Valley encephalitis virus is endemic is also likely.[12] Recently there was controversy when the chairman of the Australian National Health Medical Research Council's Public Health Committee speculated that the increase in infectious diseases sweeping Northern Australia over the previous 6 months may have been an early signal of climate change. There had been four major infectious-disease epidemics due to increased rainfall, at the hottest time of the year. A number of individuals contracted melioidosis caused by *Pseudomonas pseudomallei*, which lives in the soil and had been carried to the surface as a result of rises in the water table. An epidemic of Ross River fever had spread south, and a popular water-skiing location had been closed because of the presence of *Naegleria fowleri* (which causes amoebic meningitis). In addition, there had been outbreaks of gastroenteritis and measles. Cabinet Minister Peter Walsh attacked the claim that climate change was already established, although he agreed that if there were a warmer and more humid climate then epidemic diseases could become more troublesome and hitherto exotic diseases might become established.[13]

In the United States, five mosquito-borne diseases are potential risks after climate change—namely malaria, dengue fever and arbovirus-induced encephalitides, yellow fever, and Rift Valley fever.[14] It has been suggested in a report from the Institute of Medicine of the National Academy of Sciences that yellow fever could be a serious threat in the United States if it were to be reintroduced by a traveler infected in Africa or Brazil. The report described a hypothetical outbreak of yellow fever in New Orleans, which has a population of around 500,000. The small existing supply of vaccine would be used up within several days, and it would take some time to mobilize stocks of vaccine from Brazil. The authors predicted that 100,000 people would contract Yellow Fever and 10,000 would die within 90 days.[15]

There is good evidence of an environmental reservoir for cholera. *Vibrio cholerae* 01 has been detected on marine life in nonculturable but viable states.[16] It has been shown to coexist with a variety of algae and plankton,[17,18] which are responsive to changes in temperature, nutrient supply, and other factors, including fertilization[11] by increased concentrations of atmospheric CO_2. Climate change may influence the likelihood of cholera epidemics in the future, and it has also been proposed that the major epidemic that started in Peru in early 1991 could be the first detectable impact of global warming on health.[19] Clearly, it is impossible to be confident about this assertion at the current state of knowledge; however, the pandemics of cholera in the 19th century occurred in summer months between isotherms of 60°F and 80°F and between summer isohyets of 2 and 4 inches of rain per month.[20] Thus, if the area covered by these conditions increases or shifts, cholera may be affected accordingly. Since 1973, repeated episodes of cholera have occurred in persons living in Louisiana or Texas and in persons consuming raw oysters from Louisiana. With the rise in sea level secondary to global warming, and with changes in temperature, pH, salinity, and the composition of plant and animal life, the coastal wetlands of Louisiana and similar regions could become fertile breeding grounds for *Vibrio cholerae*.[21]

Water-borne diseases include diarrheal diseases (caused by a range of organisms, such as pathogenic *Escherichia coli,* salmonellae, and viruses), viral diseases (such as hepatitis A and poliomyelitis), giardiasis, and amoebic dysentery. They could be affected by lack of access to clean drinking water, which may occur with migration and resettlement. "Water-washed" diseases (which are made more likely by lack of water for washing) include some worm infections (such as *ascaris, trichuris,* and *oxyuris*) and many of the water-borne diseases,[11] all of which could also be affected by poor sanitation, which might be particularly likely in regions of reduced summer rainfall. Overall, childhood diarrhea would probably have the greatest impact on health.

Respiratory Diseases

Climate may affect the respiratory tract in several ways. There may be direct effects, specific weather conditions (such as thunderstorms and cold fronts), seasonal effects, and combined effects of weather conditions with other environmental or topographical factors.[22] The seasonal effects are summarized in table 2. Acute bronchitis and bronchiolitis are more common in winter in temperate countries. In addition, patients with chronic obstructive airways disease are more likely to experience exacerbations during colder months, and pneumonia is more common. By contrast, asthma and hay fever tend to peak in incidence in the summer months, and there is a second peak of acute asthma in autumn. Hospital admissions for asthma are more frequent in the last quarter of the year, probably because of asthma's association with viral infections. High levels of tropospheric ozone tend to be present in summer months and may stimulate attacks of asthma and exacerbations of chronic obstructive airways disease. They may be more likely as a result of increased UV-B flux causing increases in photochemical reactions in the atmosphere.

Thunderstorms can induce asthma attacks and have been responsible for major outbreaks. For instance, in Birmingham, U.K., in 1983 there was an approximately 20-fold increase in acute asthma seen in emergency departments over a 9-hour period. The mechanism is still unclear, but may include local buildup of allergens due to decreased

Table 2 Respiratory disease—seasonal effects. Adapted from reference 22, with permission.

	Condition affected	Possible mechanism(s)
Winter	Acute bronchitis/bronchiolitis	Colder temperature effect on bronchial responsiveness? increased viral replication/transmissibility?
	Acute exacerbations of chronic airflow obstruction/chronic bronchitis	Colder temperature increased viral replication permissive effect on secondary bacterial infection
	Asthma requiring hospital admission	Viral-induced asthma?
	Pneumonia	Colder temperature
		Associated with temperature and relative humidity 1 week before?
Summer	Asthma attacks	Pollen release
		Ozone?
	Hayfever	Pollen release
Autumn	Acute bronchitis Acute asthma	Viral infections? return to school plus reduced temperature causing increased viral transmission

winds, and a sudden drop in temperatures when the rain starts to fall, with a possible increase in positive ions. Predicting changes in respiratory disease as a result of climate change is difficult, but it is possible that increased temperatures in winter in colder countries could result in declines in bronchitis and pneumonia, while asthma and hay fever could increase. Sandstorms associated with drought could also cause localized increases in respiratory mortality in the interiors of some continents.

Implications for Food Production

The regions that seem most sensitive to the effects of climate change on food production are some semi-arid tropical and subtropical regions (including the Maghreb, the Horn of Africa, southern Africa, and eastern Brazil) and some humid tropical and equatorial regions (particularly Southeast Asia and Central America). Additionally, some regions that are currently major exporters of cereals could be affected by reduced rainfall and thus experience a decrease in production, which could have important implications for global food prices and food stocks. They include southern Europe, the southern United States, western Australia, and parts of South America.[23]

In some parts of the world, climatic change could result in the shifting of agricultural borders by 200–300 kilometers per centigrade degree of warming (equivalent to around 100 kilometers per decade under the IPCC "business as usual" scenario). While there may be some ability to adjust to such changes, it could be very limited, particularly in some marginal areas. The number of very hot days (which can cause heat stress, leading to damage to crops and to livestock) could become significantly more common in some regions. Relatively small decreases in rainfall or increases in evapo-transpiration could lead to more frequent and more severe droughts in areas already prone to such phenomena.

Although under controlled conditions higher levels of atmospheric carbon dioxide may speed the growth of some crops, such as wheat, rice, and barley, it is not clear whether the effects observed under laboratory conditions will be found under "real life" conditions. A study of a simulated rain forest suggested that plants exposed to a high concentration of carbon dioxide (610 parts per million) did not produce a greater weight of vegetation than those exposed to 340 ppm. The investigators also found increased starch grain accumulation in the leaves of the plants exposed to the high CO_2 level, which could disrupt their function. In addition, there were increased losses of CO_2 from the soil and a greater loss of nutrients in response to high levels of CO_2.[24]

Detailed studies of potential impacts on specific regions have suggested cause for concern. For instance, the implications of climate change for Canada have been studied. With an increase of 3.5°C in growing-season temperatures (assuming no change in precipitation), the frequency of drought is projected to increase 13-fold. Yields of spring wheat could be reduced by 15–37%, depending on the type of soil. As

Saskatchewan currently produces around 18% of the world's traded wheat, this reduction could have serious implications worldwide.[25]

Several studies have been undertaken to try and model the impact of climate change on the world food system. For instance, the U.S. National Center for Atmospheric Research (NCAR) suggested that, although the world's agricultural system had the capacity to absorb about two-thirds of the potential of a slowly changing climate by adjusting the land area under production, the mix of different types of crops, and the intensity of production, food stores are quite vulnerable to sudden short-term changes in crop yield. This might occur, for instance, after simultaneous droughts in major grain-producing regions. A year in which there was a 20% reduction in yield would reduce the world's reserves almost to zero.[26]

Although the proportion of the world's population receiving a diet providing less energy than 1.2 times the basal metabolic rate fell from around 23% in 1950 to 12% around 30 years later, the absolute numbers decreased only very slightly over this period because of population growth.[27] Drought caused reductions in harvest for several major producers in 1987 and 1988, and this resulted in world grain stocks' being brought to one of their lowest levels in many years. Growth in food production around the world has been limited by environmental degradation together with scarcity of crop land and water for irrigation.

Pollutants such as ozone, sulfur dioxide, and nitrous oxide may have adverse effects on crop production. In the case of ozone, for instance, damage to crops may begin to show at levels as low as 0.05 ppm for prolonged exposure of 16 days or more in the growing season. It has been calculated that during the 1980s this has led to the loss of between 5% and 10% of crops in the United States.[28]

There is already evidence of increasing death rates in a number of countries in the mid-1980s due to economic decline and perhaps to reductions in food production. In Zambia twice as many children died from malnutrition in 1984 as in 1980. In Brazil the infant mortality rate rose in 1983 and 1984, particularly in the poorest parts of the country, for the first time in decades.[29]

Climate change also has the potential to affect farm animals. For instance, it has been suggested that in North America there might be an extension of the horn fly season under some scenarios of climate change. It already causes considerable losses in the beef and dairy cattle industries. There could also be an extension of a distribution of the insect carrier of anaplasmosis, a rickettsial infection of ruminants.[30]

Currently, there is a small average net gain of 15 million tons of grain annually, but this is considerably below the 28 million tons required to keep pace with population growth. The potential effects of climate change on food security represent one of the greatest areas of concern.

Water Supply

Although it is impossible to predict exactly what will happen at a local level as a result of climate change, there is relative confidence in the types of changes that may occur at a continental level. For instance, in North America in 2020, there is predicted to be an increase in rainfall up to 20% in winter and decreases of 5–10% in summer. This could lead to a reduction in soil moisture by 10–15% during the summer months.

Nearly three-fourths of the water used currently is for agricultural purposes, although salination of irrigated land is quite widespread and climate change could further impair the capacity for development of new irrigation schemes and the maintenance of those currently existing. The water of 155 of the world's 214 first-order river systems is shared by two or more countries. Increasing international tensions over water are likely. There is already evidence of tension over the waters of the Jordan, the Euphrates, and the Mekong.[31]

Sea-Level Rise

The IPCC scientists do not foresee a rise in sea level greater than 1 meter during the next century. However, in a number of densely populated countries large numbers of relatively impoverished people are threatened by even small rises in sea level. These countries include Bangladesh, Egypt, Pakistan, Indonesia, and Thailand. Floods caused the deaths of 300,000 people in the 1970s in Bangladesh. In the Ganges

Delta in Bangladesh and the Nile Delta in Egypt, 46 million people are potentially threatened.[32] In addition, sea-level rise could cause salination of coastal areas, including estuaries and marshlands. Low-lying islands, such as the Maldives, Kiribati, and Tokelau, might no longer be habitable if sea level were to rise to the mid-range of current projections.

Other Possible Impacts

Several factors may make major disasters more likely as a consequence of climate change. There may be greater instability of climate, leading to an increased number of storms, although the Intergovernmental Panel on Climate Change has suggested that there is no clear evidence whether or not tropical or mid-latitude storms will increase. The rise in sea level combined with current patterns of storms could in itself increase the frequency of major floods. Droughts in the interior of continents could also provoke major population shifts.

A number of studies of the survivors of disasters have suggested long-term impacts on mental health. After the collapse of a dam in the United States in 1972, which killed 125 people and left 4,000 homeless, traumatic neurotic reactions were found in 80% of the survivors.[33] There were continued feelings of unresolved grief, survivor shame, and feelings of hopelessness and rage. There were developmental problems in children more than 2 years after the disaster in over 30% of cases. Other studies have suggested that psychological disturbance following disasters is of more limited duration. For example, 41% of a sample of individuals exposed to a cyclone in Darwin, Australia, showed evidence of psychological dysfunction 10 weeks after the disaster,[34] but by 14 months the level of dysfunction had returned to that of the control population. There may be increased demand for medical services for a considerable period following a disaster because of the psychological effects. In 1968 a flood in Bristol, U.K., damaged 3,000 homes. There was an increase of 53% in visits to doctors during the following year.[35] A similar impact was noted subsequent to a flood in Brisbane, Australia, in 1974.[36]

Climate change will occur against a background of other global threats, including stratospheric ozone depletion, increasing poverty, and continuing population growth in many countries. Over 90% of the projected increase in population will occur in developing countries, although in some areas AIDS will have a negative impact on population growth.[37] Altered rainfall patterns and the reduced availability of food both have the capacity to lead to conflict over basic resources. Major conflict could clearly add significantly to the burden of human suffering from climate change, with most of the impact felt by civilian populations. The proportion of civilian casualties in conflicts has risen around 50–80% since the Second World War.[38] An increasing number of refugees fleeing from environmental degradation can be perceived as threatening the security of wealthier or less affected countries. For instance, increasing numbers of migrants and refugees from North Africa are already attempting to enter Europe, and the flow of people from Central and South America to North America could increase to an even higher level than it is at present.

Conclusion

Global climate change can have direct and indirect effects on human health. The most worrying impacts are those on availability of water and food and sea-level rise, but the direct effects of temperature on disease patterns, on the vectors of communicable disease, and on the patterns of respiratory disease also give rise to concern. There is also the possibility of hitherto unpredicted outcomes, particularly in the latter half of the next century, unless temperature rise can in some way be checked. In any event, it seems clear that the potential magnitude of the effects indicates the need for major policy changes now. It has been suggested, for instance, that in order to cut annual carbon emissions worldwide to around 2 billion tons, the average per-capita emissions of carbon would have to be around one-eighth of the current Western European level by 2030.[39] This would require measures far beyond those currently being discussed at a governmental level. Health professionals can play an important role in informing the debate about the need to minimize climate change by investigating further the potential impacts of global warming on health, and by disseminating information about the potential areas of concern. The human health effects of climate change have recently been reviewed by John Last.[40]

References

1. Houghton, J. T., Jenkins, G. J., and Ephraums, J. J., eds. *Climate Change: The IPCC Scientific Assessment.* Cambridge University Press, 1990.

2. Nisbet, E. G. Some northern sources of atmospheric methane: production, history and future implications. *Canadian Journal of Earth Sciences* 26 (1989): 1603–1611.

3. Intergovernmental Panel on Climate Change. *Scientific Assessment of Climate Change.* World Meteorological Organization and UN Environmental Program, 1992.

4. Rogot, E., and Padgett, S. J. Associations of coronary and stroke mortality with temperature and snowfall in selected areas of the United States, 1962–1966. *American Journal of Epidemiology* 103 (1976): 565–575.

5. Kalkstein, L. S., et al. The impact of human induced climate warming upon human mortality; a New York case study. In Proceedings of the International Conference on Health and Environmental Effects of Ozone Modification and Climate Change, Washington, 1986.

6. Oeschli, F. W., and Buckley, R. W. Excess mortality associated with three Los Angeles September hot spells. *Environmental Research* 3 (1970): 277–284.

7. Macfarlane, A. Daily mortality and environment in English conurbations. 2. Deaths during summer hot spells in Greater London. *Environmental Research* 15 (1978): 332–341.

8. Keatinge, W. R., Coleshaw, S. R., and Eastern, J. C. Increased platelet and red cell counts, blood viscosity and plasma cholesterol levels during heat stress and mortality from coronary and cerebral thrombosis. *American Journal of Medicine* 81 (1986): 795–800.

9. Urban Climatology and Its Applications with Special Regard to Tropical Areas. Report 652, World Meteorological Organization.

10. Gillett, J. D. Direct and indirect influences of temperature on the transmission of parasites from insects to man. In *The Effects of Meteorological Factors upon Parasites,* ed. A. Taylor and R. Muller. Blackwell, 1974.

11. Potential Health Effects of Climatic Change. World Health Organization, 1990.

12. Liehne, P. F. A. Climatic influences on mosquito borne diseases in Australia. In *Greenhouse—planning for climate change,* ed. G. R. Pearman. Division of Atmospheric Research, CSIRO, Mordialloc, Victoria, Australia, 1989.

13. Lush, N. *Australia: Will greenhouse kill us? Lancet* 338 (1991): 500–501.

14. Longstreth, J. A. Human Health. In *The Potential Effects of Global Climate Change in the United States,* ed. J. B. Smith and D. Tirpak. Environmental Protection Agency, 1989.

15. Altman, L. K. "Wake-up call" issued on threats of disease. *New York Times,* October 16, 1992.

16. Huq, A. et al. Detection of *Vibrio cholerale* 01 in the aquatic environment by fluorescent–monoclonal antibody and culture methods. *Applied Environmental Microbiology* 56 (1990): 370–373.

17. Islam, M. S., Drasar, B. S., and Bradley, D. J. Long-term persistence of toxogenic *Vibrio cholerae* 01 in the mucilaginous sheath of a blue-green algae *Anabana variabilis. Journal of Tropical Medicine and Hygiene* 193 (1990): 133–139.

18. Tamplin, M. L., et al. Attachment of *Vibrio cholerae* subgroup 01 to zooplankton and phytoplankton of Bangladesh waters. *Applied Environmental Microbiology* 56 (1990): 1977–1980.

19. Epstein, P. Cholera and the environment. *Lancet* 339 (1992): 1167–1168.

20. May, J. M. *Ecology of Human Disease.* MD Publications, 1958.

21. Shope, R. Global climate change and infectious diseases. *Environmental Health Perspectives* 96 (1991): 171–174.

22. Ayres, J. G. Meteorology and respiratory disease. *Update* 40 (1990): 596–605.

23. Parry, M. *Climate Change and World Agriculture.* Earthscan, 1990.

24. Korner, C., and Arnone, J. A. Responses to elevated carbon dioxide in artificial tropical ecosystems. *Science* 257 (1992): 1672–1675.

25. Williams, G. D. V., et al. Estimating effects of climatic change on agriculture in Saskatchewan, Canada. In *The Impact of Climatic Variations on Agriculture,* volume 1: *Assessments in Cool, Temperate and Cold Regions,* ed. M. Parry, T. Carter, and N. Konijn. Kluwer, 1988.

26. Liverman, D. M. The Use of a Simulation Model in Assessing the Impact of Climate on the World Food System. Cooperative thesis 77, National Center for Atmospheric Research, 1983.

27. Grigg, D. *The World Food Problem 1950–1980.* Blackwell, 1985.

28. Brown, L., and Young, G. E. Feeding the world in the nineties. In *State of the World 1990,* ed. L. Brown. Norton, 1990.

29. UNICEF. *The State of the World's Children.* Oxford University Press, 1988.

30. Rosenzweig, C., and Daniel, M. M. Agriculture. In *The Potential Effects of Global Climate Change in the United States,* ed. J. B. Smith and J. Tirpak. Environmental Protection Agency, 1989.

31. Myers, N. Environment and security. *Foreign Policy* 74 (1989): 23–41.

32. Broadus, J., et al. Rising sea level and damming of rivers; possible effects in Egypt and Bangladesh. In *Effects of Changes in Stratospheric Ozone and Global Climate,* volume 4: *Sea Level Rise,* ed. J. G. Titus. Environmental Protection Agency and UN Environmental Programme, 1986.

33. Titchener, J. L., and Frederic, T. K. Family and character change at Buffalo Creek. *American Journal of Psychiatry* 133 (1976), no. 3: 295–299.

34. Parker, E. Cyclone Tracy and Darwin evacuees: On the restoration of the species. *British Journal of Psychiatry* 130 (1977): 548–555.

35. Bennet, G. Bristol floods 1968: Controlled survey on effects on health of local community disaster. *British Medical Journal* 3 (1973): 454–458.

36. Abrahams, M. J., et al. Brisbane floods, January 1974: Their impact on health. *Medical Journal of Australia* 2 (1976): 936–939.

37. Anderson, R. M., May, R. M., and McLean, A. R. Possible demographic consequences of AIDS in developing countries. *Nature* 332 (1988): 228–234.

38. Sivard, R. L. *World Military and Social Expenditures 1989.* World Priorities Inc., 1989.

39. Brown, L., Flavin, C., and Postel, S. Picturing a sustainable society. In *State of the World 1990.* Norton, 1990.

40. Last, J. M. Global change, ozone depletion, greenhouse warming, and public health. *Annual Review of Public Health* 14 (1993): 115–136.

Human Health and Population Growth

Michael McCally, M.D., Ph.D.

In 1950 the world contained 2.5 billion people and there was substantial but unappreciated evidence of damage to the biosphere. Now, with a world population of over 5 billion, there is inescapable evidence of environmental destruction. What is the human prospect with a world population of 7.5, or 10, or 15 billion?

Nearly 200 years ago the English economist Thomas Malthus predicted that population growth would outstrip food production, resulting in "misery and vice" and "premature death."[1] Critics then and now have argued that technology would offset population growth by increasing food production. For the last 100 years the critics have seemingly been correct and the few who warned of the dangers of population growth have largely been ignored. Evidence of health hazards and environmental damage directly attributable to population was, until recently, ambiguous.

Today there is a clear body of evidence that growth of the global population poses serious threats to human health, socioeconomic development, and the environment. But in spite of this evidence population is not an urgent public issue in developed countries. Global population growth is not a subject of public education, publicly funded research, political debate, or foreign policy. Demographers, the scientists of population, are silent or seem unconvinced of the evidence of the danger of population growth.[2] They study health, aging, and life expectancy, but they avoid questions of population density, carrying capacity, and environmental impact. Elected officials divert attention from the problem. Population was not on the agenda at the recently concluded United Nations Conference on Environment and Development in Rio de Janeiro. Agenda 21, the 800-page action prescription at Rio, contains two paragraphs on population.

The reasons for this global failure to focus on population growth are subtle and powerful:

- The political leadership of less developed countries sees the problem as one of money and development. In their view, population growth can be stabilized only by socioeconomic development financed by money from developed countries.
- Population control is seen as an imperialistic concept, deflecting attention from excessive resource consumption and pollution in developed countries.
- Some leaders count people as economic power or wealth.
- Some religions are opposed to family planning. In many cultures traditional male dominance limits the education and empowerment of women that social scientists increasingly understand to be necessary for fertility reduction. As a result, population control is a taboo subject in many parts of the world, including the United States.

A new call by scientific and political leadership for action on population issues has been issued. We are moving toward an awareness that the globe is, or soon will be, overpopulated. The 1987 Report of the World Commission on Environment and Development, *Our Common Future,* warns that "present rates of population growth cannot continue."[3] Recent emphatic calls for action on population have come from the United Nations, the U.S. National Academy of Science, the British Royal Society, former World Bank President Robert Mac-Namara, the United Nations Population Fund, the International Union for the Conservation of Nature and Natural Resources, the World Watch Institute, the Conservation Fund of the United States, and World Resources Institute, among others.

These reports come to similar conclusions. Without population control general health will remain elusive. Less developed countries will find it impossible to break out of a cycle of population growth and poverty. Without population control environmental degradation can only increase. The Malthusian prediction that population growth will finally be limited by war, pestilence, and famine seems likely to be realized.

There is a clear need for improved public understanding of population issues, for relevant research, and for appropriate public policies. Population growth has direct consequences for human health. Family planning and infant mortality reduction, which are health and medical interventions, are the basis of population-control efforts. Physicians, as scientists, as teachers, and as clinicians, have a major role in population-control efforts.

Physicians' Responsibilities

Physicians hold a special position in modern society. They are respected as healers, scientists, and authorities on matters of health. They are granted the special privileges of a profession. As a result, various responsibilities are commonly accepted by physicians, most obviously the care of patients. Because medical knowledge is important to health, physicians have traditionally held that promulgating such knowledge is one of their major responsibilities. For example, when physicians have special knowledge of dangers to health not available to others, they have a duty to inform.[4]

Physicians' responsibilities in patient care often extend to the social and political context of disease. For example, physicians regularly attempt to influence smoking cessation, seat belt use, and workplace safety. A health hazard cannot be neglected as a medical concern simply because the remedy requires political action. Although the proper form and extent of political action by physicians may sometimes be controversial, social responsibility for public health has been accepted by the medical profession since the industrial revolution.

In 1982 Christine Cassel and Andrew Jameton put forth a formal argument for the involvement of physicians in the prevention of nuclear war.[5] They reasoned that nuclear war is possible and would cause unprecedented human suffering and death, that physicians have a central moral responsibility to relieve suffering, that efforts by physicians could help prevent nuclear war, and that physicians therefore have a responsibility to work to prevent nuclear war. This conclusion was accepted by tens of thousands of physicians in the United States and hundreds of thousands of physicians around the world.

An analogous argument for environmental responsibility should compel physicians to work for the stabilization of population growth. Population growth is a health issue and clearly lies within the realm of physicians' competence and responsibility.[6]

Population Growth

Global population change over an interval of time is simply the number of births minus the number of deaths. At zero population growth the birth rate equals the death rate. For the population of a nation or a region, migration must be added or subtracted. There are a number of ways to assess population growth. Population projections use assumptions about fertility and mortality and their determinants; depending on these assumptions, low, medium, and high estimates are made of future global populations.[7,8]

Ten thousand years ago the human population numbered perhaps 100 million. At the time of Christ it was approximately 300 million. In 1850 the world's population was 1 billion. In 1930 it was 2 billion. In

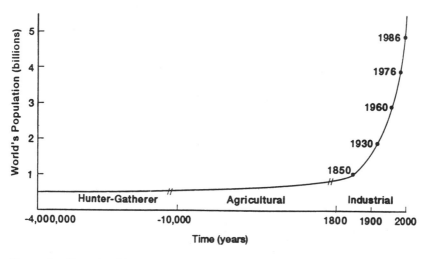

Figure 1 Growth of human population in pre-historical, agricultural, and industrial periods of world history.

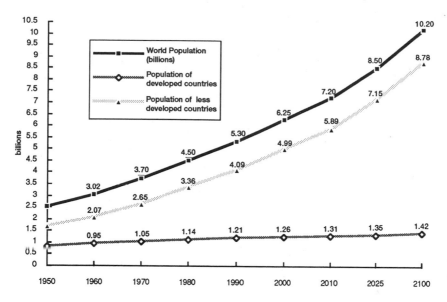

Figure 2   Growth of human population from 1900 to 2100. Adapted from ref. 12.

1990 it was 5.4 billion. The present annual growth rate of world population is about 1.8%, corresponding to a population-doubling time of about 40 years. The "rule of 69" says that in any exponential process increasing at 1% per year a population will, by compounding, double in 69 years. At present growth rates, more than 90 million persons are added to the global population each year—roughly the population of Mexico. Of this increment, 90% will occur in less developed countries, and 90% of these people will live in cities.[9]

In 1963 the United Nations' median projection suggested 6.1 billion people in the year 2000—a reasonably accurate estimate.[10] Recognizing the need for longer-range projections, in 1982 the same group made low, medium, and high projections to the year 2100.[11] This forecast was significant in that for the first time it predicted the future time at which zero population growth would be achieved. The median estimate was a final global population of 10.2 billion, while the high-growth estimate was 15 billion at year 2100 (still short of zero population growth). These projections assumed a maximum life expectancy of 75 years for men and 80 for women. High rates of contraceptive use are also assumed in the low-growth projections.

The United Nations reassessed its projections in 1990. The United Nations Population Fund now states that the world's population will reach 10.0 billion in 2050 and will stabilize in 2150 at 11.6 billion, as opposed to the 10.2 billion estimated a decade earlier.[12] The report suggests a worst-case scenario, with little use of family planning, in which world's population would quadruple to 20 billion. The report attributes 80% of the world's deforestation to population growth. Other projections give alarmingly higher steady-state predictions.

The Demographic Transition

With the Industrial Revolution, improvements in the socioeconomic determinants of health caused death rates to fall; population increased accordingly. Later, birth rates in industrial nations declined and their populations stabilized. This complex process is termed the *demographic transition*.[7,13]

The demographic transition describes population growth experienced by countries during industrialization. In the first stage, both birth and death rates are high but similar, so population growth is low. In the second stage, industrialization causes a "population explosion": the mortality rate, particularly that for infants, falls in response to socioeconomic improvements. Population growth is, therefore, rapid. In the last stage, the fertility rate also falls and the population stabilizes, but at a higher level. The demographic transition model is a useful way to think about population growth, but it has been reformulated in light of more recent understandings.[14,15]

The Demographic Trap

Lester Brown noted that in many less developed countries a fall in mortality is not followed by a fall in fertility.[16] He coined the term *demographic trap* for the circumstance in which the fertility rate does not fall in the second stage of the transition and a high rate of population growth is sustained. When this happens, the rapid population growth prevents the developing country from achieving the economic and social gains necessary to limit birth rates and to slow the growth of the population.

Population Demography

The rate of annual global population increase reached a peak of 2.04% between 1965 and 1970. It declined to 1.7% by 1989, but since then it has increased to 1.8% (reflecting, perhaps, the recent increase in birth rates in China). The low-growth countries of Europe, North America, and the Pacific are growing at only 0.5%; the less developed countries are growing at 2.4% a year.

These growth rates seem innocuous, but the population problem is with the absolute numbers. The world's population has been increasing steadily for a long time. Only recently have we heard of global warming, acid precipitation, loss of species, and desertification. Whereas population growth is smooth and predictable, environmental change may be abrupt and nonlinear.[17] Consider the example of a community of fishermen around a lake.[17] The population of fishermen increases without having a serious impact on the fish population until a threshold or crucial point is reached at which the reproduction of the fish is interfered with and the fish population collapses. Between 1900 and 1950 sardines were in huge abundance in California's Monterey Bay and supported a large industry. "Fished out," they disappeared suddenly, and they have not returned.

Population scientists also make an important distinction between renewable and nonrenewable resources. For the near term there seem to be sufficient nonrenewable resources.[18] The prices paid for most metals, for natural gas, and for oil have declined or remained low in the last half-century. In the United States gasoline is cheaper than bottled water. Ironically, it is the renewable resources—lumber, water, and fish, for example—that in our increasing numbers we are exhausting.

Births and Deaths

Birth rates are declining in nearly all countries.[19] In 1960 births per woman worldwide were approximately six. Today it is closer to three. But the absolute number of births continues to increase and will not peak, according to United Nations estimates, until sometime in the middle of the 21st century. Death rates are also declining. But while globally we now have 140 million births a year, we only have 50 million deaths.[20] As a result, we are adding each year more than 90 million persons—the population of Mexico—to the world's population.

Infant mortality has fallen substantially around the world in the last four decades, most dramatically in the less-developed countries. Deaths of infants under 1 year of age per 1,000 births fell in developed countries from 56 to 10, and in less developed countries from 180 to 56.[20]

Population Momentum

Zero population growth is an elusive goal. If, for example, we had reduced global fertility rates to replacement levels in 1990, when the population was 5.3 billion, population growth would nonetheless continue until the year 2050, reaching roughly 7 billion. The reason for this "momentum" lies in the age and sex composition of populations in developing countries (figure 4). Because less developed countries have a large proportion of young people and few older persons, a baby boom will occur even at low fertility rates as these young people move through their reproductive years. Population growth continues for several generations after fertility is reduced to replacement levels.[21]

Overpopulation and Food

It is highly unlikely that food supplies can be increased enough to be adequate for a doubled global population. The world's grain reserves have been falling since 1988. The current supply would last approximately 60 days. There may be insufficient food to feed the numbers expected by the end of the 1990s.[22]

Agricultural intensification can go some way toward maintaining a balance between food production and population, but there are limits to this solution. Productivity gains from intensive agriculture are limited by availability of energy and fertilizer, water supplies, and the global loss of topsoils. In estimates assuming low levels of technology, 64 countries now lack the resources to feed themselves.[22] To meet the food needs of a world population of 10 billion would require massive international redistribution of resources.[3,23]

Desertification

In developing countries there is a population-driven phenomenon of land degradation that scientists term *desertification*. Desertification is the impoverishment of land by human activities.[24] Its causes are overgrazing, overcultivation, salinization, and deforestation. Desertification occurs when the carrying capacity—the number of people a given area of land can support—is exceeded. It turns out that firewood resources may be the limiting factor for carrying capacity in some temperate and hot climates. Natural cycles of drought further lower a region's carrying capacity and hasten desertification. According to a 1984 United Nations assessment, 4.5 billion hectares (35% of the earth's land surface) are threatened by desertification. Figure 3 shows World Bank estimates of the loss of forest in Africa at three levels of projected population growth. This land sustains 20% of the world's population.[3] Global warming would intensify these processes.

Overgrazing of marginal lands by cattle is a major cause of desertification. Environmentalists in the United States have recently proposed reform of the livestock economy, with the goal of significant reductions in the use of meat as a foodstuff.[25] A campaign called "Beyond Beef" seeks a 50% reduction in beef consumption in the next 10 years, arguing that cattle husbandry contributes needlessly to rainforest destruction, global warming, water pollution, water scarcity, desertification, and world hunger.[26]

Technology vs. Population

The belief that population growth is an important cause of environmental deterioration is not universally held. It has also been argued that environmental impact is not correlated with population growth and that affluence and polluting technologies are responsible.[27] Both positions are probably correct, population being the proximate cause and technology being the ultimate cause of environmental degradation.[28] Technology is destroying the environment in the developed countries, while human populations are desertifying the developing world. Developed countries must limit their affluence and technology, while less developed countries must constrain their numbers. We must all worry about both.

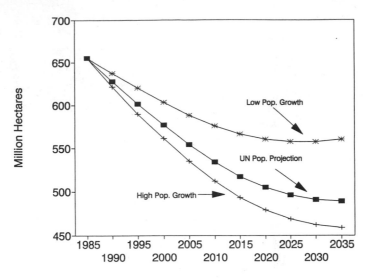

Figure 3   Total forest area and future population growth in Africa from 1985 to 2035. Source: David Wheeler, Environmental Indicators Program, World Bank.

Population-Related Health Issues

Teen Pregnancy

Adolescent pregnancy is a major population issue in both developed and less developed countries. Age at the time of a first child is an important determinant of lifetime fertility, birth interval, and family size. The annual rate of 95 pregnancies per 1,000 women aged 15–19 in the United States is among the highest in the world. The rate in Canada is 46; in the Netherlands the rate is 15.[8] Adolescent pregnancies are at high risk of low birth weight and preventable infant mortality. Prevention of adolescent pregnancy is a major unsolved population health issue in the United States.

Maternal Mortality

A major public health problem associated with high fertility rates is maternal mortality, one of the great neglected problems of health care in developing countries. The World Health Organization estimates that approximately 500,000 women die each year from causes related to pregnancy.[29] Of these deaths, 95% occur in developing countries, where maternal mortality rates are as much as 100 times higher than in industrialized countries. The causes of this mortality are infection, obstructed labor and ruptured uterus, eclampsia, hemorrhage, and the complications of induced abortion.[30]

The same efforts (improved access to health and family-planning services) that are needed to lower fertility rates in developing countries will lower pregnancy-related maternal mortality. Family planning in these cases is both life-saving and fertility-lowering.

Health Effects of Urbanization

The movement of people to cities is a hallmark of population change in the 20th century. In 1900 only 12–14% of persons lived in cities. By the year 2000, half of us will live in cities, with the greatest urban growth occurring in less developed countries.

Urbanization involves changes in population density, housing, nutrition, sanitation, air and water quality, and access to health services, each of which may have separate and important effects.[31,32] In cross-sectional studies, urban dwellers as a whole have distinct socioeconomic advantages over rural populations. In a recent study in Nigeria, urban adults were more literate, earned higher wages, and had more access to clean water, toilets, and health-care facilities than their rural counterparts. The city dwellers also had lower infant mortality and increased life expectancy.[33] Urbanization can have a beneficial effect on the health of populations in the short term.

However, among recent urban immigrants living in barrios the situation is reversed. In the Nigerian study, high infant mortality was associated with overcrowding and pollution.[33] In Mexico City there are so many people without sanitary facilities that a "fecal snow" often falls on the city as the wind picks up dried human waste.[34]

Population Density

Coupled with poverty and poor sanitation, crowding contributes to epidemic infectious illness. But population density alone means little for health. In Africa there are 55 persons per square mile, in Europe 261, and in Japan 857—numbers inversely related to population health status.

The important concept is carrying capacity. A country or a region may be considered overpopulated when it exceeds the capacity of its environment to support the population. In this sense, Japan and Europe are overpopulated, as they must draw on many other environments than their own for resources. The social consequences of overcrowding in communities of caged animals have been described,[35] but social scientists have been reluctant to extrapolate these studies to human society.

Refugees

There are between 15 million and 50 million refugees from civil unrest, wars, and natural disasters worldwide. Some number of these leave their homes fleeing famine and desertification; these persons have been called "ecological refugees."[23] Disaster and relief resources are increasingly inadequate to meet the well-known health needs of these new refugees.

Population Aging

As the mortality rate declines in most parts of the world, life expectancy and the number of older persons increase. Although the phenomenon of population aging is most characteristic of developed countries, it is, nevertheless, seen worldwide.[36] The epidemiological transition is a three-phase movement from high mortality (due to epidemic infectious illness and famine) through a period of lengthening life expectancy and on to a third phase of chronic and degenerative illness, with heart disease, cancer, and stroke predominating as causes of death.[37] The services required to care for the chronically ill create new economic burdens, particularly in less developed nations. As fewer young persons are present to support their elders, the dependency ratio increases. This phenomenon may create political pressure for increases in fertility if governments cannot provide for elders. The government of Japan has expressed concern about that country's rising dependency ratio.

Conditions Necessary for Population to Stabilize

There are four health-related conditions that seem necessary to limit population growth: adult literacy, low infant mortality, access to family-planning services, and improvements in women's status. These issues

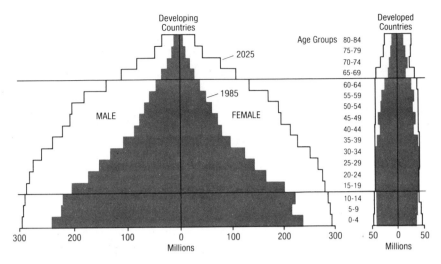

Figure 4 Population age pyramids for 1985 and 2025 in developing and developed countries. Source: ref. 12.

must be addressed simultaneously, in a sustained and culturally appropriate fashion. Population growth is a complex of causes and effects and must be considered holistically; leaders may initiate programs, but persons change their reproductive behaviors only from their own motivations.

One of the most interesting case studies of the demographic transition in a developing country was seen in Kerala, a province in southwestern India. Population growth was stabilized even though per-capita incomes remained extremely low. Provincial leaders, using international funding, developed a plan that reflected Kerala's social, religious, and political character. They achieved high rates of literacy, lowered infant mortality, and made birth-control services easily accessible to achieve the remarkable result of near-zero population growth in a poor country.[21]

Literacy

Literacy among women is a remarkably powerful predictor of birth rate. Women's ability to control their own reproductive behaviors is strongly associated with their literacy. Literacy seems to be more than a correlate of income, years in school, or employment; independent of these factors, it is positively associated with low birth rates.[38]

Infant Mortality

Belief that your offspring will die young is an incentive to have many children. Low infant mortality rates encourage the belief among parents that, even with a small family, some of their children will survive to maturity, carry the family name, help with subsistence tasks, produce family income, and provide for the parents when they are old. The African leader Julius Nyerere noted several decades ago that "the most powerful contraceptive is the confidence of parents that their children will survive."

Some people argue that infant-mortality-reduction programs without simultaneous efforts in family planning and birth-rate reduction are unethical. Maurice King, a distinguished international health physician, has argued that saving the lives of infants in undeveloped countries condemns them to more suffering later on and "increases the man-years of human misery."[39-41] The ethical dilemma is resolved by the recognition that both family planning and primary health-care services for infant-mortality reduction are necessary for birth-rate reduction and must be part of an integrated approach to women's health.[42,43]

Primary health-care services and family planning services are organized, funded, and delivered by separate agencies in most parts of the developing world. This historical situation has created bureaucracies that compete for funds and do not see their programs as related.

Family Planning

Family planning, declared a basic health need by the United Nations and the World Health Organization, includes the right of individuals and couples "freely and responsibly to decide the number and spacing of their children and to have the information, education, and means to do so."[29] Women who lack effective family planning often face an unfortunate choice between unwanted pregnancy and abortion.

Family planning works; it lowers birth rates. Access to birth control has increased over the last three decades, but there still is a large unmet need for contraception and safe abortion. Worldwide, 43% of married couples are controlling their fertility through the use of modern contraceptives. In the developed world 52% of couples are contraceptive users, and in China the rate is 73%. In the developing world only 27% use contraceptives. Only a small percentage of the population of Africa is known to have access to family-planning services.[44,45]

Family planning is political. In 1984, in response to pressure from anti-abortion groups in the United States, the Reagan administration stunned the UN-sponsored International Conference on Population in Mexico City by declaring that foreign countries were not eligible for U.S. family planning assistance if they used their own funds for abortion services, counseling, or referral. Similarly, in 1991 a "gag rule" was placed on federally funded (Title X) domestic family-planning programs in the United States. The "Mexico City" policy led to the withdrawal of U.S. funds from the London-based International Planned Parenthood Federation. In 1986, the United States withdrew from the UN Population Fund in protest of China's compulsory birth-control programs.

In the absence of accessible and universally effective contraception, women use abortion as an alternative method of preventing or spacing pregnancies.[46] Opponents of abortion argue that it should not be a method of family planning. In 1973, the same year that the Supreme Court decriminalized abortion in the United States, Congress passed the Helms Amendment to the Foreign Assistance Act, prohibiting the use of foreign-aid funds for abortion services in recipient countries. The United States will not contribute to family-planning programs that include abortion activities, irrespective of legality. Whatever their personal feelings about abortion, most physicians believe that safe and up-to-date contraception is essential to women's health and will reduce reliance on abortion for birth control.

Women's Health and Abortion

Perhaps 30 million legal abortions and 10–22 million clandestine abortions were performed worldwide in 1987.[47] Others estimate there are 40–50 million abortions per year, of which one-quarter to one-third are illegal or clandestine.[48] Legal-abortion rates range from a high of 112 per 1,000 women of reproductive age per year in the Soviet Union (before its breakup) to a low of 5 per 1,000 in the Netherlands. The rates in the United States and Canada were 28 and 10, respectively. Abortion rates, while increasing in some countries and declining in others, have been relatively constant around the world in recent decades. The worldwide trend in the last decade has been toward the liberalization of abortion laws.[49]

Public health officials estimate that between 100,000 and 200,000 women die each year in developing countries from the complications of clandestine abortion. Rumania provides an example of the relation of population policy and abortion. Abortion policies were liberalized in 1957, and the birth rate declined. In 1966, under the bizarrely pronatalist policies of President Ceaucescu, abortion became illegal and the birth rate rose. By 1983 the birth rate had again declined, because of clandestine contraception and a high rate of illegal abortion. The liberalization of abortion laws after the revolution in 1989 produced a prompt fall in maternal mortality, which had been the highest in Europe.

Research in contraception and reproductive health has been sharply limited in the United States in the last decade. Only one major pharmaceutical company is actively developing new contraceptives. The controversy surrounding research and development of RU 486, the anti-progestin "abortion pill," and the French manufacturer's refusal to license the drug in the United States, are clearly due to pressure from anti-abortion organizations. (A recent clinical study found that RU-486, used with prostaglandin as a first-trimester abortifacient, had a 96% efficacy rate.[50]) Clinical research on progesterone (Norplant) implants has been similarly delayed.

Control of population growth appears to be in women's hands. The empowerment of women, meaning access to education, health services, employment and public life, is coming to be understood as a major determinant of fertility. These observations have led one writer to suggest that "a male dominated society is a threat to public health."[51]

Population Policies

The receipt of foreign aid is critical to countries struggling to control their populations. But contributions from developed countries are paltry. In the early 1960s the World Council of Churches recommended that the foreign aid contributed by the wealthy nations should be 1% of their gross national product (GNP). In 1970 this standard was lowered by the United Nations to 0.7%. Few nations have reached even this standard. In 1990 the donor countries' average aid contribution was 0.35% of the GNP. The United States' foreign-aid spending in that year was 0.21% of the GNP.[51] Only a small portion of development assistance, about 5%, is for health-related activities, including family planning. A sensible policy for global population control would return to the 1% standard, with much larger fractions for family-planning services and women's health.

Population issues are complex and must be addressed by each nation and its citizens. The complexities include cultural and religious attitudes toward sexuality and the role of women, funds for development, and power relationships between developed and less developed countries. In his book *Earth in the Balance,* Vice-President Albert Gore proposed a "global Marshall Plan" for preservation of the planet.[52] To ensure population stabilization, he recommended that resources be developed for literacy, general education, and public health programs to reduce infant mortality. Furthermore, Gore asserted, access to contraception should be made universal, with culturally appropriate instruction and with a particular emphasis on breast feeding.

Sound policy initiatives recommended by numerous groups and agencies include universal access to family planning, development assistance at 1% of the GNP of developed countries, broadened education and full economic and political participation of women, public and private support for research in birth-control technologies, mass communication aimed at increasing support for family planning, and the integration of family planning and primary health-care services for women and children.

Actions Physicians Can Take

Man has been said to have a biological inability to perceive slow changes in his environment.[53] Man is neurophysiologically organized to respond to rapidly changing sensory inputs and to ignore slow changes. A frog that is dropped into hot water jumps out, but one that is very slowly heated in cool water will not perceive the temperature change and will boil. In like fashion, man appears not to perceive slow changes in population and in the physical environment—a partial explanation for our failure to take action.

Individual physicians need to become population activists. Activism is an inclusive term covering everything from letter writing and awareness to running for political office. It can be a full-time job, as in running a non-profit community-service organization. It can also become integrated into one's professional life. Physicians are in an ideal position to educate others about the risks of population growth and environmental damage.

The development of national and international networks of concerned health professionals is a high priority.[54] In the United States and Canada, Physicians for Social Responsibility organized doctors to educate the public and political leaders about the medical consequences of global nuclear war. International Physicians for the Prevention of Nuclear War won the 1985 Nobel Peace Prize for these activities. A similar coalition of health workers from around the world is now needed to educate the health professions, alert the public to the health consequences of population growth and environmental degradation, and create policy.

This work had begun. Physician environmental groups have come into being in at least a dozen countries, including Switzerland, Austria, Germany, Italy, Australia, New Zealand, Canada, Sweden, and Belarus. An International Society of Doctors for the Environment was founded in Italy in 1990.[55] Many national affiliates of IPPNW are active in global environmental issues. Regrettably, population is not a major issue for these organizations.

Attention was brought to these issues at a conference on Human Health and the Environment held in Cambridge, Massachusetts, on October 10 and 11, 1992, under the sponsorship of Physicians for Social Responsibility, the United Nations Environment Programme, the medical schools of Brown and Tufts Universities, International Physicians for the Prevention of Nuclear War, the Sierra Club, the Natural Resources Defense Council, the Union of Concerned Scientists, the Massachusetts Institute of Technology, and the Harvard School of Public Health.[56] The organizers believe that this event will stimulate the organization of physicians in the United States around issues of environment and population.

References

1. Malthus, T. *An Essay on the Principle of Population* (1798). Penguin, 1985.

2. Ohlin, G. The population concern. *Ambio* 31 (1992), no. 1: 6–9.

3. World Commission on Environment and Development. *Our Common Future.* Oxford University Press, 1987.

4. Jonsen, A. R., and Jameton, A. J. Social and political responsibilities of physicians. *Journal of Medical Philosophy* 2 (1977): 376–400.

5. Cassel, C., and Jameton, A. Medical responsibility and thermonuclear war. *Annals of Internal Medicine* 97 (1982): 426–432.

6. McCally, M., and Cassel, C. K. Medical responsibility and global environmental change. *Annals of Internal Medicine* 113 (1990): 467–473.

7. Coleman, D., and Schofield, R. *The State of Population Theory: Forward from Malthus.* Blackwell, 1986.

8. Last, J. M. The demographic trap and sustainable health. *Canadian Journal of Public Health* 82 (1990): 3–4.

9. El-Badry, M. A. World population change: A long-range perspective. *Ambio* 21 (1992), no. 1: 18–23.

10. World Population Prospects. Population study 41, Population Division, Department of Economic and Social Affairs, United Nations, 1963.

11. Population bulletin 14-1982, Population Division, Department of International Economic and Social Affairs, United Nations, 1982.

12. Sadik, N. *The State of World Population 1991.* United Nations Population Fund, 1991.

13. Notestein, F. W. Population—The Long View. In *Food of the World*, ed. T. Schultz. University of Chicago Press, 1945.

14. Caldwell, J. Toward a restatement of demographic transition theory. *Population Development Review* 21 (1976), no. 3–4: 321–366.

15. Keyfitz, N. Completing the worldwide demographic transition: The relevance of past experience. *Ambio* 21 (1992), no. 1: 26–30.

16. Brown, L. Analyzing the demographic trap. In *State of the World 1987*, ed. L. Brown.

17. Keyfitz, N. Population growth can slow the development that would slow population growth. In *Preserving the Global Environment: the Challenge of Shared Leadership*, ed. J. T. Mathews. Norton, 1991.

18. Simon, J. *Population Matters: People, Resources, Environment and Immigration.* Transaction Publishers, 1990.

19. Kunitz, S. Mortality since Malthus. In *The State of Population Theory: Forward since Malthus,* ed. D. Coleman and R. Schofield. Blackwell, 1986.

20. Hill, K. Fertility and mortality trends in the developing world. *Ambio* 21 (1991), no. 1: 79–83.

21. Keyfitz, N. The growing human population. *Scientific American* 26 (1989), no. 3: 119–126.

22. Brown, L. Food. In *State of the World 1992,* ed. L. Brown. Norton, 1992.

23. Brown, L. Reexamining the world food prospect. In *State of the World 1989,* ed. L. Brown. Norton, 1989.

24. Postel, S. Halting land degradation. In *State of the World 1989,* ed. L. Brown. Norton, 1989.

25. Durning, A. T., and Brough, H. B. Reforming the livestock economy. In *State of the World 1992,* ed. L. Brown. Norton, 1992.

26. Rifkin, J. *Beyond Beef: The Rise and Fall of the Cattle Culture.* Dutton, 1992.

27. Commoner, B. *Making Peace with the Planet.* Pantheon, 1990.

28. Shaw, R. P. Rapid population growth and environmental degradation: Ultimate versus proximate factors. *Environmental Conservation* 16 (1989), no. 3: 199–208.

29. Royston, E., and Armstrong, S. *Preventing Maternal Deaths.* World Health Organization, 1989.

30. Rosenfeld, A. Maternal mortality in developing countries: An ongoing but neglected "epidemic." *Journal of the American Medical Association* 262 (1989): 376–379.

31. Bourne, P. *Water and Sanitation: Economic and Social Perspectives.* Academic Press, 1984.

32. Monteiro, C. A., and Benicio, M. H. Determinants of infant mortality trends in developing countries—some evidence from the San Paulo city. *Transactions of the Royal Society of Tropical Medicine and Hygiene* 83 (1989): 5–9.

33. Williams, B. Assessing the health impact of urbanization. *World Health Statistics Quarterly* 43 (1990): 145–152.

34. Ehrlich, P. R., and Ehrlich, A. E. *The Population Explosion.* Simon and Schuster, 1990.

35. Calhoun, J. B. Population and social pathology. *Scientific American* 140 (1962): 139–148.

36. Olshansky, J., and Cassel, C. K. In search of Methuselah: Estimating the upper limits to human longevity. *Science* 250 (1990): 634–640.

37. Kjelstrom, T., and Rosenstock, L. The role of environmental and occupational hazards in the adult health transition. *World Health Statistics Quarterly* 43 (1990): 188–196.

38. Grosse, R. N., and Auffrey, C. Literacy and health status in developing countries. *Annual Review of Public Health* 10 (1989): 281–297.

39. King, M. Health is a sustainable state. *Lancet* 336 (1990): 664–667.

40. King, M. Public health and the ethics of sustainability. *Tropical Geography Medicine* 42 (1990): 197–206.

41. Jolly, R. Overpopulation and death in childhood. *Lancet* 336 (1990): 936.

42. Booth, R. *Homo sapiens*—A species too successful. *Journal of the Royal Society of Medicine* 83 (1990): 757–759.

43. Waldman, R. J. Child survival and global sustainability: Malthus revisited. *Physicians for Social Responsibility Quarterly* 1 (1991), no. 4: 177–184.

44. *A Report on Progress Towards Population Stability.* Population Crisis Committee on Access to Affordable Contraception, 1991.

45. Fathalla, M. F. Family planning: Future needs. *Ambio* 21 (1992), no. 1: 84–87.

46. Dixon-Mueller, R. Abortion policy and womens' health in developing countries. *International Journal of Health Sciences* 20 (1990), no. 2: 297–314.

47. Henshaw, S. K. Induced abortion: A world review, 1990. *Family Planning Perspectives* 22 (1990), no. 2: 76–89.

48. Jacobsen, J. L. Coming to grips with abortion. In *State of the World 1991*, ed. L. Brown. Norton, 1991.

49. Kunin, H., and Rosenfeld, A. Abortion: A legal and public health perspective *Annual Review of Public Health* 12 (1991): 361–382.

50. Silvestre, L., et al. Voluntary interruption of pregnancy with Mifepristone (RU 486) and prostaglandin analogue. *New England Journal of Medicine* 322 (1990): 645–648.

51. Mann, J. Reverence for life and community solidarity: An international perspective. *Physicians for Social Responsibility Quarterly* 2 (1992), no. 1: 51–56.

52. Gore, A. *Earth in the Balance.* Houghton Mifflin, 1992.

53. Ornstein, R., and Ehrlich, P. *New World, New Mind: Moving Towards Conscious Evolution.* Doubleday, 1989.

54. Last, J. M. *Textbook of Public Health.* Oxford University Press, 1992.

55. Nussbaumer, W. International Society of Doctors for the Environment. In *Doctors for the Environment.* ISDE Scientific Office, 1991.

56. Epstein, P. R. Health and the environment. *Lancet* 340 (1992): 1030–1031.

Species Extinction and Biodiversity Loss: The Implications for Human Health

Eric Chivian, M.D.

The loss of diversity is the most important process of environmental change . . . because it is the only process that is wholly irreversible.

—E. O. Wilson[1]

. . . one species has become so efficient at reproducing itself and dominating all other forms of life that it is in the act of endangering all species, including itself.

—D. E. Koshland[2]

Learn from the beasts the physic of the fields.

—Alexander Pope

Human activity is causing the extinction of animal, plant, and microbial species at rates which are thousands of times those that would have occurred naturally[3] and which approximate those of the largest extinctions in geological history. When *Homo sapiens* evolved, some 100,000 years ago, the number of species on earth was the largest ever.[1] Current rates of species loss are reducing the number of species to the lowest since the end of the age of dinosaurs, 65 million years ago. It is estimated that one-fourth of all species will become extinct in the next 50 years.[4] (A species is best defined as a population of organisms in which all the physiologically competent individuals are capable of breeding under natural conditions with all the other individuals of the opposite sex in that population.[5])

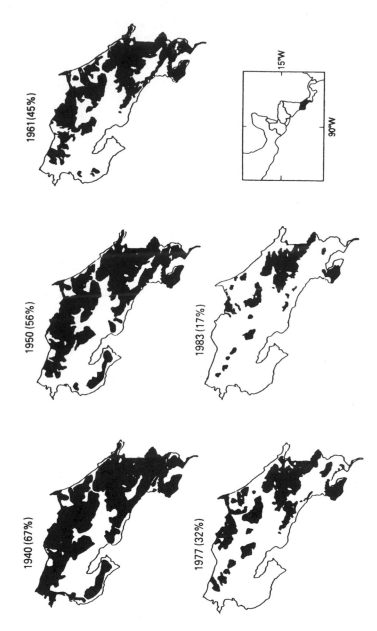

Figure 1 Deforestation in Costa Rica, 1940–1983. Percentages refer to the country's land area covered by primary forests. By 1983 only rugged montane rainforests remained relatively undisturbed. Adapted from reference 7 with permission from Oxford University Press and *Biotropica*.

In addition to the ethical issue that we have no right to kill off countless other organisms, many of which came into being tens of millions of years prior to our arrival, this behavior is ultimately self-destructive. It upsets the delicate ecological balance on which all life depends, including our own, and it destroys the biological diversity that makes soils fertile, creates the air we breathe, and provides food and other life-sustaining natural products, most of which remain to be discovered.

The exponential growth in the human population (there are 100 times more of us than of any land animal of comparable size that ever existed[3]), the even greater rise in the consumption of resources (we consume or destroy 20–40% of all the solar energy trapped by land plants[3]), and the production of wastes are the main factors endangering the survival of other species. Global warming, acid rain, the depletion of stratospheric ozone, and the discharge of toxic chemicals into the air, the soil, and freshwater and marine ecosystems all ultimately lead to a loss of biodiversity.

But it is habitat destruction by human activities, particularly deforestation, that is the greatest destroyer. This is especially the case for tropical rainforests. Less than 50% of the area covered by prehistoric tropical rainforests remains,[3] but these forests are still being cut and burned at a rate of approximately 142,000 square kilometers per year (about 1.8% of the total amount still standing[6]). This loss is equal in area to Switzerland and the Netherlands combined. Each second's loss of forest cover is the size of a football field.[3] It is this destruction that is primarily responsible for the mass extinction of the world's species.

The diversity of life in these forests is extraordinary. For example, one investigator found an average of 300 distinct tree species in a series of 1-hectare plots in Peru.[8] (The total number of native species of trees in all of North America is 700!) E. O. Wilson discovered 43 species of ants (from 26 genera) on a single tree in these same Peruvian forests[9]— about the same number of ant species as in all of the British Isles. This enormous biological diversity has led many to conclude that tropical rainforests, which comprise only 6% of the world's land mass, contain perhaps 50% of the world's species.

Estimates of the total number of species vary widely. Approximately 1.5 million species have been recorded and given scientific names.[10] Two-thirds of these are in temperate regions, and most of them are insects. But given the enormous diversity of life in tropical rainforests, and based on figures extrapolated from species counts of canopy-dwelling beetles in Panama (dropped by an insecticidal fog), Erwin[11] and others[4,12] have estimated that there are somewhere between 10 million and 100 million different species on Earth. If we count viruses, many of which adapt to only one or two species (for example, there are more than 100 different viruses adapted almost exclusively to humans[13]), even these estimates may be low.

New species are constantly being discovered, both in the tropics (the *Maues* marmoset, for example, is the third new primate species to be found in the Amazon since 1990[14]) and elsewhere (e.g. at great depths in the oceans[15]), and the known species of bacteria, protozoans, helminths (flatworms), and mites most likely represent only a very small fraction of the total. Bacteria, for instance, are unbelievably diverse. Only 4,000 species have been cataloged worldwide, yet in a single gram of soil from a beech forest Norwegian investigators found 4000–5000 different bacterial species[3] (and they found a similar number, with little overlap, in a gram of sediment from off the coast of Norway[3]), almost none of which had been previously described. It is clear from these examples that we are in a state of almost total ignorance about the enormous variety of life on Earth, particularly in the case of microscopic organisms.

Even if a conservative estimate of 20 million total world species is used (which may be off by a factor of 5 or more), then 10 million species would be found in tropical rainforests, and at current rates of tropical deforestation this would mean that 27,000 species would be lost in tropical rainforests alone each year—more than 74 per day, 3 per hour.[3] Many species in these rainforests are confined to single, isolated habitats[16] whose destruction would wipe them out, as they would not be able to relocate to adjacent undisturbed areas; this suggests that even the above figures may be underestimates.

If people fully comprehended the effect these massive species extinctions will have—in foreclosing the possibility of understanding and treating many incurable diseases, and ultimately, perhaps, in threatening human survival—then they would recognize that the current loss of biological diversity represents nothing less than a medical emergency, and would demand that efforts to preserve species and ecosystems be given the highest priority.

The Loss of Medical Models

It would seem difficult to believe that species so different from our own, such as frogs, insects, corals, forest vines, and bacteria, could offer us much in the way of understanding human physiology and illness. Yet on a biochemical level, and especially on a genetic level, we are all made from the same mold. All life contains ribonucleic acids as its primary information code and uses similar chemical mechanisms to regulate metabolism, to reproduce, and to defend itself against invading organisms. Even when separated by millions of years of evolution, as we are from the chimpanzees, species are strikingly alike—our DNA is almost identical to that possessed by a chimp, differing in composition by only 1%.[17]

It is because of these similarities that valuable models for study can be found among other species—models that offer windows for understanding ourselves. Without them, medicine would be lost.

Consider Gregor Mendel's peas, which revealed the basic principles of heredity, or the fruit fly (*Drosophila melanogaster*), whose large chromosomes have become familiar to every high school biology student in the world. Were it not for these "lowly" organisms, and others like them (such as the bacteria *Escherichia coli*, which dwells symbiotically in our intestines), the biology of genetics, and our ability to understand various human hereditary conditions, would be in the Dark Ages.

Three groups of endangered animals, far apart in the animal kingdom—dart-poison frogs, bears, and sharks—offer striking examples of how models that are important for biomedical science are in danger of being squandered by man.

Dart-Poison Frogs

The entire family of dart-poison frogs (the *Dendrobatidae*) is threatened by destruction of its habitats—the lowland tropical rainforests of Central and South America.[18] These brightly colored frogs, which include more than 100 species, are particularly sensitive to deforestation, as they often live only in very specific areas of the forest and cannot live naturally anywhere else. Through the work of Charles Myers, John Daly, and Edson Albuquerque,[19,20] who are rushing to collect and study these unique animals before they disappear, we have come to understand that the toxins they produce—used for centuries to poison arrows and blow-gun darts by Central and South American Indians—are among the deadliest natural substances known. They are also enormously useful to medicine.

The active ingredients of the frog toxins are alkaloids—nitrogen-containing ring compounds found almost exclusively in plants (morphine, caffeine, nicotine, and cocaine are examples). Frog alkaloids bind selectively to specific ion channels and pumps in nerve and muscle membranes (sodium and potassium channels, calcium pumps, and acetylcholine receptors[20]). Without them, our knowledge of these basic units of membrane function, found throughout the animal kingdom, would be very incomplete.

For example, batrachotoxin, one of the most potent of all the toxins produced by dart-poison frogs, binds to a specific site in sodium channels[20]; it has been critically important in defining the action of this fundamental molecular structure. (Surprisingly, this same deadly toxin has been found in a completely different organism thousands of miles from the dart-poison frogs—it coats the feathers and the skin of the Hooded Pitohui bird of New Guinea.[21])

In addition to their value in basic neurophysiological research, dart-poison frogs offer valuable biochemical clues for the production of new and potent analgesics that have a mechanism of action different from that of morphine, of new medicines for cardiac arrhythmias, and of new treatments for the alleviation of some neurological diseases (such as Alzheimer's disease, myasthenia gravis, and amyotrophic lateral sclerosis).[18] If rainforest destruction continues at its present rate in Central and South America, these extremely valuable frogs will be lost—and, because the structure of their alkaloids is too complex for laboratory synthesis to be practical,[18] their medically important toxins will be lost as well.

Bears

Asia's growing black-market trade in bear parts, with gall bladders being sold for their reputed medicinal value (at 18 times their weight in gold, more than the price of the best heroin) and paws for gourmet food,[22] coupled with continued hunting and the destruction of habitats, has imperiled bear populations in many parts of the world. If some species of bears become extinct, we will all be the poorer, not only because they are beautiful, fascinating creatures that fill important ecological niches, but also because some species possess two unique physiological processes that may ultimately provide important clues for the treatment of osteoporosis and kidney failure.

Hibernating black bears are immobile for up to 5 months in the winter, yet do not lose bone mass.[23] Whereas a human would lose almost one-fourth of his or her bone mass during a similar period of immobility (or lack of weight bearing), a bear continues to lay down new bone, making use of calcium circulating in its blood.[23] Understanding the mechanisms of how bears accomplish this feat may result in our being able to prevent osteoporosis in the elderly (an extremely widespread and costly health problem, leading to fractures, pain, and disability), in those confined to bed for long periods, and in astronauts subjected to prolonged weightlessness.

In addition, a hibernating bear does not urinate for months. Humans who cannot excrete their waste products in urine for several days build up high levels of urea in their blood and die from its toxicity. Somehow bears recycle their urea to make new proteins, including those in muscle.[23] If we could determine the mechanism of this process, it might lead to effective, long-term treatments for persons with kidney failure, who must now rely on the frequent use of dialysis machines.

Sharks

Like bears, many species of sharks are being wiped out because of the demand for shark meat—especially in Asia, where shark fins for soup command prices as high as $100 a pound.[24] Because sharks produce few offspring, grow slowly, and take years to mature, they are highly vulnerable to overfishing.

Sharks have been around for almost 400 million years and have evolved highly specialized organs and physiological functions that have protected them against virtually all threats—except slaughter by man. The decimation of shark populations and the extinction of some of the 350 species are likely to constitute a major disaster for human beings.

As the "apex" predator of the oceans, sharks are critical to the maintenance of marine ecosystems. Their absence could trigger severe disruptions in the balance of other populations, with marked effects on marine food supplies. In addition, sharks may have the most advanced sensory systems of all the vertebrates,[24] with highly sophisticated and sensitive auditory, visual, vibratory, "smell," and electric field receptors, and a large and complex brain. Understanding these unique sensory physiologies may provide insights into our own "more primitive" systems and suggest ways for assisting persons with sensory impairments.

The immune systems of sharks (and of their relatives, skates and rays) seem to have evolved so that the animals are almost invulnerable to cancers and infections. Tumors, often seen in other fishes and in mollusks,[25] are rare in sharks. Preliminary investigations have supported this finding. For example, it has proved impossible to produce tumor growth in nurse sharks with repeated injections of known potent carcinogens,[24] and MIT researchers have isolated from Basking Shark cartilage[26] a substance (probably similar to the protein they identified in bovine cartilage[27]) that strongly inhibits the growth of new blood vessels toward solid tumors, and thereby prevents tumor growth.

Sharks may also provide valuable models for developing new types of medications to treat infections. Squalamine, a compound recently isolated from the tissues of the dogfish shark, has demonstrated potent activity in laboratory tests against a variety of bacteria, fungi, and parasites through mechanisms not yet understood. A steroid, chemically similar to cholesterol but different from all other known classes of antibiotics, squalamine may lead to new antibiotics for infectious organisms that have become resistant to standard drugs.[109]

Other Models

Other toxins and chemical compounds found in a wide variety of animals, some of which are endangered, have also shown great promise as research models for understanding human physiology and disease:

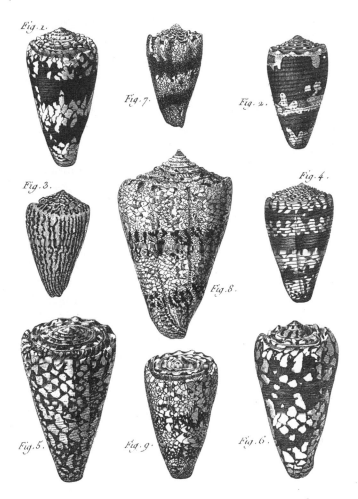

Figure 2   Some species of cone snails. These coral reef mollusks contain a wide variety of toxins useful to medical research. Source: reference 28.

- Conus toxins are small peptides produced by predatory cone snails (a large genus of venomous snails found in tropical reef communities). Each of the approximately 500 species may produce a different type or assortment of venoms.[29] These toxins are so numerous as to rival in diversity the alkaloids of plants and the secondary metabolites of microorganisms (which supply us with many of our most important antibiotics). Conus toxins interact with a wide variety of channels and receptors in neuromuscular systems, including calcium and sodium channels and NMDA (N-methyl-D aspartate), acetylcholine, and vasopressin receptors.[29] Most of their channel and receptor activity has yet to be characterized. Several conus toxins are now standard tools in neurophysiological research, and some are used in clinical medicine (for example, as a diagnostic test for myasthenia gravis).

- Pseudopterosins are highly active substances from the Caribbean sea whip, another reef creature. They possess potent anti-inflammatory and analgesic properties, the mechanisms of which have not been defined.[30] More potent than currently available nonsteroidal anti-inflammatory agents, such as indomethacin, these substances may prove invaluable in understanding and treating cellular responses that produce pain and inflammation.

- The prostaglandins are lipid compounds that have been shown to mediate an enormous range of physiological responses in human beings, including gastrointestinal motility and secretions, blood flow, endocrine secretions, and the inflammatory response.[31] But research has been hampered by the limited availability of prostaglandins—they occur in only very minute quantities in terrestrial organisms. In recent years, however, prostaglandins have been discovered in sea creatures in very large quantities. For example, up to 8% of the wet body weight of the Caribbean gorgonian coral *Plexaura homomalla,* a soft coral which is extremely plentiful in some coral reef communities, is prostaglandin A2.[32] These abundant new sources have revolutionized the field of prostaglandin research.

Coral reefs, the "rainforests of the seas," and their resident organisms (such as cone snails, sea whips, and soft corals), are increasingly threatened with destruction worldwide[33] as a result of overfishing, of increased seawater temperatures, and of pollution from agriculture, industry, and sewage.

- Tetrodotoxin, another selective binding compound for sodium channels, is produced by the puffer fish,[34,35] the ocean sunfish, and the porcupine fish.[36]

- Charybdotoxins, which specifically block the action of potassium channels, are found in some species of scorpions.[37]

- In the 1960s, it was found that the venom from pit vipers contained compounds that enhanced the response to bradykinin, a potent vasodilating agent, by blocking its destruction. Subsequent work led to a greater understanding of the role of the renin-angiotensin system in the maintenance of blood pressure, and to the development of an entirely new class of extremely effective and safe medications to treat hypertension: the angiotensin converting enzyme (ACE) inhibitors.[38] Examples are captopril and lisinopril.

Countless other examples could be mentioned of known unique plants, animals, and microorganisms (and there are likely to be millions more still unknown), holding the secrets of billions of evolutionary experiments, that are increasingly threatened by human activity and in danger of being lost forever to medical science.

The Loss of New Medicines

Plant, animal, and microbial species are themselves the sources for some of today's most important medicines and make up a significant proportion of the total pharmaceutical armamentarium. Of all prescriptions dispensed from community pharmacies in the United States from 1959 through 1980, 25% contained active ingredients extracted from higher plants[39]; the percentage is much higher in the developing world. As many as 80% of all people living in developing countries—roughly two-thirds of the world's population—rely almost exclusively on traditional medicine using natural substances, most of them derived from plants.

The knowledge held by traditional healers, often passed down orally over centuries, has led to the discovery of many medicines that are widely used today—quinine, physostigmine, d-tubocurarine, pilocarpine, and ephedrine, to name a few.[40] But that knowledge is fast disappearing, particularly in the Amazon,[41,42] as native healers die out and are replaced by more modern medical practitioners. Botanists and pharmacologists are racing to learn these ancient practices, which, like the forest plants they employ, are also endangered.[39,41,42]

In this section, I shall briefly review some of the most useful natural medicines, mention others that hold promise, and discuss how the extinction of species and the loss of biodiversity may affect pharmacological medicine.

Figure 3   *Cinchona officianalis,* the South American tree that produces quinine and quinidine. Source: reference 43.

Quinine and Quinidine

Quinine was one of the most important finds in the New World by Europeans. Extracted from the bark of cinchona trees by South American Indians, who used it to treat fevers, it was shown to be highly effective in curing malaria, still one of the most widespread and deadly diseases of the tropics.[44] Quinine has been largely replaced by synthetic analogues, but it is making a comeback as resistance has begun to develop to the more widely used synthetic drugs.[45]

Some patients who were being treated for malaria with cinchona extracts, and who also had atrial fibrillation, were found to have converted to normal sinus rhythm.[46] Subsequent investigations established that it was the d-isomer of quinine, quinidine, that was most effective in this process. Today quinidine is one of the most widely used drugs in the treatment of cardiac arrhymias, both atrial and ventricular.

D-Tubocurarine

Curare is a generic term for a group of chemicals, derived from South American vines of the genus *Chondrodendron,* that are used by South American Indians as arrow poisons. After the paralyzing properties of the curares become known to explorers, the curares became the object of intensive research, which led to their being used as treatments for tetanus muscles spasms and other spastic disorders (1932) and as muscle relaxants during general surgery (1942).[47] The most active agent in the *Chondrodendron* extract was d-tubocurarine, and this compound and its synthetic derivatives have become some of today's most important drugs. Almost all general surgery relies on these neuromuscular blocking agents to achieve deep muscle relaxation (especially important during abdominal and orthopedic operations) without high doses of general anesthetics.

Vinblastine and Vincristine

While exploring claims by practitioners that the tropical rosy periwinkle plant (*Vinca rosea*) could be used to treat diabetes mellitus, R. L. Noble observed that his treated laboratory rats had granulocytopenia and bone-marrow suppression.[48] Subsequent studies with a variety of tumors in test animals established that rosy periwinkle extracts were highly effective in preventing replication in tumor cells, and led to the purification of two vinca alkaloids that showed significant anti-tumor activity: vinblastine and vincristine.

Vinblastine is effective against testicular tumors, Kaposi's sarcoma (seen in AIDS), neuroblastoma, and cancer of the breast in women, but is particularly effective against Hodgkin's Disease, achieving marked improvement in 50–90% of cases.[48] Vincristine has been shown to be the most effective agent known against acute childhood leukemias, increasing the remission rate from 20% to 90%.[48] Massive deforestation in Madagascar has threatened the rosy periwinkle, leading to its cultivation in large plantations.

Streptomycin

The above pharmaceuticals are all derived from higher plants found in tropical forests. However, microorganisms from tropical soils are also fertile sources of medicines, and are also lost through deforestation. Stimulated by the discovery of penicillin from bread mold, S. Waksman and his co-workers conducted a systematic search of fungi from tropical soils during the period 1939–1943.[49] Their search led to powerful antibiotics from fungi of the *Streptomyces* genus—most importantly streptomycin, which was shown to inhibit growth of the tubercle bacillus and which became a major treatment for tuberculosis and for other infectious diseases, including streptococcal bacterial endocarditis and plague. These discoveries subsequently led to a generation of broad-spectrum antibiotics and antifungal agents from other tropical soil fungi—neomycin, amphotericin, chloramphenicol and erythromycin, to name a few.

All of the above are derived from tropical organisms, and it is in the tropics—particularly in the rainforests—that the greatest diversity of life can be found. But temperate regions are also vast storehouses of medicines, about which extremely little is known. A few of the most important will be mentioned below.

Aspirin

Hippocrates noted that the ancient Greeks used a brew of willow leaves as an analgesic in childbirth,[50] but not until the middle of the 18th century was it discovered that the bark of the willow tree (*Salix alba*) could lower fevers.[51] The active ingredient, salicin, was isolated during the 19th century, and was made more palatable as acetylsalicylic acid (now synthesized in the laboratory) by Felix Hoffman at the German pharmaceutical firm Bayer as the century came to a close. Americans consume 16,000 tons of aspirin per year for pain relief (particularly in muscles and joints and for headaches), for fevers, as anti-inflammatory

---

agents (especially for rheumatoid arthritis), and more recently as a preventive against thrombotic myocardial infarctions and strokes. The exact mechanism by which aspirin achieves all these wonders has remained obscure, but in the last 20 years research has shown that aspirin's main effect is the blocking of prostaglandin synthesis.[51]

Digitalis

Digitalis from the foxglove plant (*Digitalis purpurea*), an herb popular in gardens for its beautiful flowers, was used by Welsh physicians in the 13th century. Its effectiveness in treating peripheral edema ("dropsy") and in stimulating the heart was not described until the 18th century.[52] Today, digitalis (as digoxin and digitoxin) is widely used to increase the contractility of cardiac muscle in heart failure, and to treat atrial flutter and fibrillation as well as paroxysmal atrial and AV nodal tachycardia.

Morphine

Opium, derived from the opium poppy (*Papaver somniferum*), has been used medicinally since at least the third century B.C.[53] Morphine, the most active alkaloid in opium along with other active compounds (e.g., codeine and papaverine) and their synthetic derivatives, remains among the most powerful and useful drugs in all of medicine—as an analgesic, as a sedative and anti-anxiety agent, and as a treatment for diarrhea, cough, and dyspnea from acute left ventricular failure. In addition, the use of these compounds has led to the discovery of naturally occurring opioid compounds and receptors in the human brain and other organs, and to the elucidation of their many functions.

Psoralens

Psoralens, compounds found in citrus fruits, parsley, celery, figs, the subtropical *Psoralea corylifolia,* and the Umbelliferae plant (*Ammi majus*), have been used in India in combination with exposure to sunlight since the 14th century B.C. to treat skin depigmentation. Used topically or taken by mouth, psoralens are thought to work by increasing both the proliferation of melanocytes (the melanin-producing cells in skin) and the production of melanin itself. In recent years, both naturally occurring compounds (methoxsalen) and their synthetic derivatives (trioxsalen) have been widely used with ultraviolet A exposure for severe cases of psoriasis. There is also evidence that psoralens are effective against cutaneous T-cell lymphomas.[110]

Taxol

The drug taxol, derived from the bark of the Pacific yew tree (*Taxus brevifolia*) in old-growth forests of the Pacific Northwest, is one of the success stories of the National Cancer Institute's screening program that has been looking for plants, animals, and microbes with anti-cancer activity. Discovered in 1962,[55] taxol has been shown to be highly effective in unresponsive cases of advanced ovarian cancer,[56] with 30% of patients having significant remissions. Currently the most effective treatment available for this disease, it has been called "the most promising cancer-fighting discovery in 15 years."[57] It has also been found to be effective in metastatic breast cancer[55] and in acute leukemias.[58]

It is estimated that four mature yew trees (it takes 100 years to reach full maturity) are needed to produce enough taxol to treat a single case of ovarian cancer. However, "clear cutting" of forests in the Pacific Northwest has squandered the yew supply,[59] and synthesis of the extremely complicated taxol molecule is not thought likely to become commercially feasible.[55] There are indications that other yew species also contain taxol, as well as taxol analogues with anti-cancer potential. And *Taxomyces andreanae,* a fungus associated with the Pacific yew, has been shown to produce taxol independently.[108]

Ricin

Ricin, a compound from the seeds of the castor bean plant (*Ricinus communis*), has been of interest for centuries to those contemplating homicide, because of its extreme toxicity to man in quantities difficult to detect.[60] In recent years, this highly potent cytotoxic agent (it takes only one or a few molecules to kill most cells[60])—a glycoprotein that inhibits protein synthesis by inactivating ribosomal subunits—has been a key factor in the development of immunotoxins, antibody-toxin compounds that are extremely useful in bone-marrow transplants and in treating autoimmune diseases but are particularly important in the treatment of cancers.[61]

Figure 4   *Taxus brevifolia* (the Pacific yew), source of taxol. Illustration by Suzan Strobel.

Immunotoxins are the present-day equivalent of Paul Ehrlich's "magic bullet."[62] The principle is that a molecule that specifically recognizes and binds to a target cell is joined to another that is highly toxic to that cell, resulting, in the ideal case, in the selective destruction of the target cell. Ricin is currently being joined to monoclonal antibodies that are developed from, and bind to, specific tumor cells.

It is widely believed that immunotoxins using ricin, and other cytotoxic agents derived from plants, fungi, and bacteria, will lead to significant new advances in cancer chemotherapy.[63]

Figure 5   *Taxus brevifolia*. Source: reference 54.

Cytarabine

The oceans, despite being filled with countless highly evolved organisms containing bioactive compounds likely to yield useful medicines, have produced only one drug in current use: cytarabine, also known as ara–C or cytosine arabinoside (the Upjohn trade name is Cytosar-U).

Sponges (along with sharks and tunicates) rarely contain tumors,[25] perhaps as a consequence of their synthesis of substances that inhibit carcinogenesis and/or tumor growth. The discovery of large amounts of arabinosyl nucleosides, compounds with marked anti-metabolite activity, in the Caribbean sponge (*Tethya crypta*)[36] seemed to offer an explanation for the absence of tumors in this sponge species, and led to the synthesis of cytarabine. Arabinosyl nucleosides are pyrimidine analogs, compounds that inhibit DNA synthesis or RNA function.[48]

Cytarabine is the most effective drug for inducing remission in acute myelocytic leukemia in children and adults.[48] In combination with other cancer chemotherapeutic agents, it is also useful in the treatment of non-Hodgkin's lymphomas and acute lymphocytic leukemias.

Potential New Medicines

An active search for new pharmaceuticals is underway. The National Cancer Institute and several of the major pharmaceutical houses (including Merck and Bristol-Meyers Squibb) are leading the field, and new companies and researchers are entering the market, with strategies that employ ethnobotanical knowledge. One company that actively researches medicines used by native South American healers, Shaman Pharmaceuticals, has reported a promising new drug that appears to be effective against the influenza A and B viruses when used systemically, and against Herpes Simplex I and II viruses (even when resistant to acylovir) when used topically.[64]

Other highly promising medicines obtained from plants and animals include potent new anticoagulants from vampire bats[65] and from leeches, blackflies, and ticks[66]; antiviral compounds from fireflies[67]; an antimalarial from the Chinese herb *Artemisia anua*[68]; unique anti-inflammatory agents from marine sponges[69]; and potent anti-tumor medicines from mosses,[70] marine tunicates,[71] and bryozoans.[25]

Scientists have analyzed the chemistry of fewer than 1% of the known rainforest plants for biologically active substances[72] (and of a similar proportion of temperate plants[73]), and of even smaller percentages of the known animals, fungi, and microbes. But there may be tens of millions of undiscovered species in forests, in soils, and in lakes and oceans. With the massive extinctions currently in progress, we may also be destroying new cures for incurable cancers, for AIDS, for arteriosclerotic heart disease, and for other illnesses that cause enormous human suffering.

In a similar manner, ticks, tsetse flies (which carry African sleeping sickness), sylvan mosquitoes, and kissing bugs (which carry Chagas disease, an endemic disease in many areas of Central and South America that affects 15–20 million people and is the leading cause of heart disease in these zones[79]) can, in the short term, infect people who destroy their forest habitats and drive out their usual hosts.

- In areas where there is little biological diversity, such as on small islands and in deserts, the risk of human exposure to vector-borne diseases can be higher than in areas where there is great diversity. This is true because vectors are often indiscriminate in their habits, and when there is a greater variety of animals there are more targets for attack.[77] Examples are the greater risk of getting Lyme disease on Nantucket Island (where there is a paucity of fauna) than on the mainland, or of getting leishmaniasis while camping in the Sinai Desert.[78] If the biological diversity of animals becomes markedly reduced in an area, it is presumed that the same principle will apply.

- When natural landscapes are greatly altered and their ecosystems damaged by human activity, foreign insect species, usually introduced by accident, often thrive better than native species.[80] Consider the Asian tiger mosquito, *Aedes albopictus,* which came to the United States in water-filled used tires and is now multiplying rapidly, being better adapted than native mosquitoes, in damaged environments in 21 states. *Aedes albopictus* has been demonstrated to carry eastern equine encephalitis (EEE), a viral encephalitis with a mortality of 30%, in Polk County, Florida,[81] and there is concern that the mosquito could precipitate epidemics of EEE in areas where the virus is already endemic. *Aedes albopictus* is also a competent vector for dengue fever and yellow fever. Although its opportunistic feeding habits make it less of a danger to man than its cousin *Aedes aegypti* (the prime dengue and yellow fever vector, which feeds almost exclusively on human hosts), it could still pose a major threat to public health in the United States, especially if increases in precipitation and temperature from global warming extend its range (along with that of *Aedes aegypti*) to more northerly latitudes.[82]

- The proliferation of vectors can also result in outbreaks of infectious diseases. The massive losses in the bat[75] and bird[76] populations worldwide, for example, may have enormous consequences for the spread of insect-borne diseases.

Other conditions can support increases in vector populations as well. For example, cutting oak or maple forests and leaving the stumps results in the formation of sprout-encircled water basins that provide ideal breeding grounds for the mosquito *Aedes triseriatus,* which carries La Crosse encephalitis.[83] This widespread viral illness, which accounted for 5–6% of all acute central nervous system disease in the midwestern region of the United States from 1966 through 1986[84] (and which is also found in the northeastern states), is chiefly seen in boys between the ages of 5 and 10. Although it carries a relatively low mortality (2%), it can result in much higher percentages of residual seizures and behavioral problems.[84]

- When reservoir hosts multiply, one can see a spread of infectious diseases. For example, the unprecedented increase in the deer populations in North America (due in part to the destruction of deer predators but also, and perhaps more importantly, to reforestation) has led to a burgeoning of the deer-feeding tick population and to outbreaks of Lyme disease in exposed populations.[77] Lyme disease, an illness caused by the bacterial spirochete *Borrelia burgdorferi,* can present as a mild, short-lived, flu-like syndrome with a characteristic rash, or, if untreated, as a chronic illness with arthritic and neurologic complications.[85]

- The history of malaria offers important lessons about the effects of human activity on the spread of vector-borne diseases. It has been hypothesized that malaria initially became a major public health problem several thousand years ago in western Africa when early "slash and burn" agriculture began to encroach upon the rainforests.[86] The newly cleared areas favored the breeding of the malaria-carrying mosquito *Anopheles gambiae,* which fed mostly on human beings, over the more omnivorous mosquito species that were less able to adapt to the new environments. According to Stephen Morse, "by disrupting the established ecological order, people inadvertently encouraged the adaptation of a 'weed' species."[86]

More recently, in Brazil, malaria has reached epidemic proportions as a consequence of massive settlement and environmental disruption of the Amazon basin. Largely under control in Brazil during the 1960s, malaria exploded 20 years later, with 560,000 cases reported in 1988 (500,000 in Amazonia alone[87]). In large part, this epidemic was a consequence of the influx of huge numbers of people who had little or no immunity to malaria, who lived in makeshift shelters, and who wore little protective clothing. But it was also an outgrowth of their disturbing the environment of the rainforest, creating in their wake stagnant pools of water everywhere—from road construction, from silt runoff secondary to land clearing, and from open mining. In these pools, *Anopheles darlingi,* the most important malaria vector in the area, could multiply unchecked.[87]

- The story of "emerging" viral illnesses may hold valuable clues for understanding the effects of habitat destruction on human beings. For example, Argentine hemorrhagic fever, a painful hemorrhagic viral disease with a mortality rate between 3% and 15%,[88] has been epidemic since 1958 as a result of the widespread clearing of the pampas of central Argentina and the planting of corn.[86] The virus is carried by the native mouse *Calomys musculinus,* which has adapted to and flourished in the cornfields. As the original biologically diverse grasslands have been increasingly replaced by corn monoculture, Argentine hemorrhagic fever has spread dramatically, from an area of 6,000 square miles in 1955 to one of 40,000 square miles in 1990.[86]

Another viral infection with hemorrhagic manifestations that has recently "emerged" is Kyasanur Forest disease, a sometimes prolonged systemic illness with a mortality rate of about 5% that is carried by *Haemaphysalis spinigera* ticks in the tropical forests of Mysore in southern India.[84] The spread of this disease is thought to result from deforestation and the introduction of sheep grazing into the cleared areas, favoring a proliferation of the tick vectors.[89]

In both of these cases, the mechanisms seem clear: the disease vectors, whose numbers were held in check by various natural ecosystem balances, proliferated in the new environments to which they were better adapted than their competitors, and spread their viruses to man. But the "emerging" viral illness that has had the greatest impact on human health, and which may be a harbinger of future viral outbreaks, is AIDS, caused by the human immunodeficiency viruses HIV-1 and HIV-2. There is general agreement that the current AIDS epidemic originated from nonhuman primates in Africa, which have acted as natural, asymptomatic hosts and reservoirs for a family of immunodeficiency viruses.[90] Good genetic evidence exists for the links of HIV-1 to a simian inmmunodeficiency virus in African chimpanzees[91] and of HIV-2 to another simian virus in African sooty mangabeys.[92,93] Are these cross-species viral transmissions between nonhuman primates and man due to man's encroachment into degraded forest environments? If this is the case, we may be witnessing with AIDS the beginning of a series of viral epidemics originating from tropical rainforests,[13] where there may be tens of thousands of viruses that could infect humans. Some of these viruses may be as lethal as AIDS (approaching 100%) but may spread more easily (for instance, by airborne droplets). These potential viral diseases could become the most serious public health consequence of the destruction of the rainforests.

Other Effects

The disruption of other relationships among organisms, ecosystems, and the global environment, about which almost nothing is known, may prove the most catastrophic of all for human beings. For example, what will happen to the global climate and to the concentration of atmospheric gases when some critical threshold of deforestation has been reached? Forests ("the Earth's lungs") play a crucial role in the maintenance of global precipitation patterns and in the stability of atmospheric gases.

What will be the effects on marine life if increased ultraviolet radiation causes massive ocean phytoplankton kills, particularly in the rich seas beneath the Antarctic ozone "hole"? These organisms, which are at the base of the entire marine food chain, are highly vulnerable to ultraviolet damage.[94–96]

What will be the consequences for plant growth if acid rain and toxic chemicals poison soil fungi and bacteria that are essential for soil fertility? There has already been a 40–50% loss of species of fungi in Western Europe during the past 60 years (including many symbiotic mycorhizal fungi,[3] which are crucial to the absorption of nutrients by plants). No one understands what the effects of this loss will be.

Scientists do not know the answers to these and other critically important questions, but worrisome biological signals suggest that major damage to global ecosystems has already occurred. The rapid simultaneous disappearance of many species of frogs worldwide, even in pristine environments far from people, indicates that they may be dying as a consequence of airborne pollutants carried over long distances or from some other global environmental change, such as increased ultraviolet radiation.[97] The immediate cause of death, in many of the cases, seems to be opportunistic infections (the amphibian equivalents of the kinds of infections seen in AIDS patients), secondary to a collapse of the frogs' immune systems.[98]

Closer to humans, marine mammals (such as striped dolphins in the Mediterranean, European seals off the coast of Northern Ireland, and beluga whales in the St. Lawrence River) are also dying in record numbers. In the case of the dolphins and the seals, some of the deaths seem to be due to infections by morbilli viruses causing pneumonias and encephalitides[99,105] (perhaps also the consequence of compromised immune systems). In the case of the whales, chemical pollutants such as DDT, Mirex, PCBs, lead, and mercury seem to be suppressing fertility and causing a variety of tumors and pneumonias.[100] The beluga whales' carcasses were often so filled with these pollutants that they could be classified as hazardous waste.

Are these "indicator species," like canaries that die in coal mines containing poisonous gases, warning us that we are upsetting fragile ecosystem balances that support all life, including our own? The 50% drop in sperm counts in healthy men worldwide during the period 1938–1990,[101] the marked increases in the rate of congenital malformations of the external genitalia in males in England and Wales from 1964 through 1983,[102] the dramatic rise in some cancer incidence rates for white children in the United States from 1973 through 1988,[103] and the steady growth in the mortality rates for several cancers worldwide for the last 3 to 4 decades[104–107] all suggest that environmental degradation may be starting to compromise not only the survival of frogs, marine mammals, and other animal, plant, and microbial species, but that of the human species as well.

Summary

Human activity is causing the extinction of animal, plant, and microbial organisms at rates that may well eliminate one-fourth of all species on Earth within the next 50 years. The incalculable human health consequences from this destruction include the loss of medical models of human physiology and illness; the loss of new medicines that may successfully treat incurable cancers, AIDS, and other diseases causing great human suffering; and the upsetting of the balance among ecosystems on which all life, including human life, depends. Major efforts to protect natural habitats and to preserve biodiversity are required to prevent these medical catastrophes from occurring.

References

1. Wilson, E. O. Threats to biodiversity. *Scientific American* 261 (1989), no. 3: 108–116.

2. Koshland, D. E. Preserving biodiversity. *Science* 253 (1991): 717.

3. Wilson, E. O. *The Diversity of Life*. Harvard University Press, 1992.

4. Ehrlich, P. R., and Wilson, E. O. Biodiversity studies: Science and policy. *Science* 253 (1991): 758–762.

5. Wilson, E. O. The current state of biological diversity. In *Biodiversity*, ed. E. O. Wilson. National Academy Press, 1988.

6. Myers, N. *Deforestation Rates in Tropical Forests and Their Climatic Implications.* Friends of the Earth, 1989.

7. Whitmore, T. C. *An Introduction to Tropical Rainforests*. Clarendon, 1990.

8. Gentry, A. H. Tree species richness of upper Amazonian forests. *Proceedings of the National Academy of Sciences* 85 (1988): 156–159.

9. Wilson, E. O. *Biotropica* 19 (1987): 245.

10. May, R. M. How many species are there on earth? *Science* 241 (1988): 1441–1449.

11. Erwin, T. L. *Bulletin of the Entomological Society of America* 30 (1983): 14.

12. Stork, N. E. *Biological Journal of the Linnaean Society* 35 (1988): 321.

13. Preston, R. Crisis in the hot zone. *New Yorker,* October 26, 1992.

14. New type of monkey is found. *New York Times,* October 20, 1992.

15. Yoon, C. K. In dark seas biologists sight a riot in life. *New York Times,* June 2, 1992.

16. Diamond, J. M. *Discover* 11 (1990): 55.

17. Ackerman, D. Insect love. *New Yorker,* August 17, 1992.

18. Brody, J. E. Using the toxin from tiny frogs, researchers seek clues to disease. *New York Times,* January 23, 1990.

19. Albuquerque, E. X., Daly, J. W., and Withop, B. Batrachotoxin: Chemistry and pharmacology. *Science* 172 (1971): 995–1002.

20. Myers, C. W., and Daly, J. W. Dart-poison frogs. *Scientific American* 248 (1983), no. 2: 120–133.

21. Dumbacher, J. P., et al. Homobatrachotoxin in the genus *Pitohui*: Chemical defense in birds. *Science* 258 (1992): 799–801.

22. Montgomery, S. Grisly trade imperils world's bears. *Boston Globe,* March 2, 1992.

23. Rosenthal, E. Hibernating bears emerge with hints about human ills. *New York Times,* April 21, 1992.

24. Stevens, W. K. Terror of the deep faces harsher predator. *New York Times,* December 8, 1992.

25. Tucker, J. B. Drugs from the sea spark renewed interest. *BioScience* 35 (1985), no. 9: 541–545.

26. Lee, A., and Langer, R. Shark cartilage contains inhibitors of tumor angiogenesis. *Science* 221 (1983): 1185–1187.

27. Moses, M. A., Sudhalter, J., and Langer, R. Identification of an inhibitor of neovascularization from cartilage. *Science* 248 (1990): 1408–1410.

28. Lamarck, J. P. *Tableau Encyclopédique et Méthodique des trois règues de la Nature. Vers coquilles, mollusques et polypiers,* Tôme Troisième. Mme. Veuve Agasse, Paris, 1827.

29. Olivera, B. M., et al. Diversity of conus neuropeptides. *Science* 249 (1990): 257–263.

30. Look, S. A., et al. The pseudopterosins: Anti-inflammatory and analgesic natural products from the sea whip *Pseudopterogorgia elisbethae. Proocedings of the National Academy of Sciences* 83 (1986): 6238–6240.

31. Campbell, W. B. Lipid-derived autocoids: Eicosanoids and platelet-activating factor. In *The Pharmacological Basis of Therapeutics,* eighth edition, ed. A. G. Gilman et al. Pergamon, 1990.

32. Gerhart, D. J. Prostaglandin A2: An agent of chemical defense in the Caribbean gorgonian *Plexaura homomalla. Marine Ecology—Progress Series* 19 (1984): 181–187.

33. Williams, E., and Bunkley-Williams, L. The world-wide coral reef bleaching cycle and related sources of coral mortality. *Atoll research Bulletin* 335 (1990): 1–71.

34. Fuhrman, F. A. Tetrodotoxin, tarichatoxin, and chiriquitoxin: Historical perspectives. In *Tetrodotoxin, Saxitoxin, and the Molecular Biology of the Sodium Channel,* ed. C. Y. Kao and S. R. Levinson, *Annals of the New York Academy of Sciences* 479 (1986): 1–14.

35. Lange, W. R. Puffer fish poisoning. *American Family Physician* 42 (1990), no. 4: 1029–1033.

36. Ruggieri, G. D. Drugs from the sea. *Science* 194 (1976): 491–497.

37. Potter, D. Personal communication, December 9, 1991.

38. Garrison, J. C., and Peach, M. J. Renin and angiotensin. In *The Pharmacological Basis of Therapeutics,* eighth edition, ed. A. G. Gilman et al. Pergamon, 1990.

39. Farnsworth, N. R. The role of ethnopharmacology in drug development. In Bioactive Compounds from Plants, Ciba Foundation Symposium, 1990.

40. Farnsworth, N. R., et al. Medicinal plants in therapy. *Bulletin of the World Health Organization* 63 (1985), no. 6: 965–981.

41. Schultes, R. E., Dwindling forest medicinal plants of the Amazon. *Harvard Medical Alumni Bulletin,* summer 1991: 32–36.

42. Balick, M. J. Ethnobotany and the identification of therapeutic agents from the rainforest. In *Bioactive Compounds from Plants,* Ciba Foundation Symposium, 1990.

43. Humboldt, A. von, and Bonpland, A. *Plantae aequinoctiales.* Lutetiae Parisiorum, Paris, 1808–1809.

44. WHO Task Group on Potential Health Effects of Climatic Change World Health Organization, 1990.

45. Webster, L. T. Drugs used in the chemotherapy of protozoal infections: Malaria. In *The Pharmacologic Basis of Therapeutics,* eighth edition, ed. A. G. Gilman et al. Pergamon, 1990.

46. Bigger, J. T., and Hoffman, B. F. Antiarrhymic drugs. In *The Pharmacologic Basis of Therapeutics* (op. cit.).

47. Taylor, P. Agents acting at the neuromuscular junction and autonomic ganglia. In *The Pharmacologic Basis of Therapeutics* (op. cit.)

48. Calabresi, B. A., and Chabner, B. A. Antineoplastic agents. In *The Pharmacologic Basis of Therapeutics* (op. cit.).

49. Sande, M. A., and Mandell, G. L. Antimicrobial agents: The aminoglycosides. In *The Pharmacologic Basis of Therapeutics* (op. cit.).

50. Saltus, R. New uses for aspirin give it star billing. *Boston Globe,* December 30, 1991.

51. Weissmann, G. Aspirin. *Scientific American,* January 1991: 84–90.

52. Hoffman, B. F., and Bigger, J. T. Digitalis and allied cardiac glycosides. In *The Pharmacologic Basis of Therapeutics* (op. cit.).

53. Jaffe, J. H., and Martin, W. R. Opioid analgesics and antagonists. In *The Pharmacologic Basis of Therapeutics* (op. cit.).

54. Sargent, C. S. *Silva of North America,* volume 10. Houghton Mifflin, 1896.

55. Daly, D. Tree of life. *Audubon,* March–April 1992: 76–85.

56. McGuire, W. P., et al. Taxol: A unique antineoplastic agent with significant activity in advanced ovarian epithelial neoplasms. *Annals of Internal Medicine* 111 (1989), no. 4: 273–279.

57. FDA panel backs taxol as cancer drug. *New York Times,* November 17, 1992.

58. Rowinsky, E. K., et al. Phase I and pharmacodynamic study of taxol in refractory acute leukemias. *Cancer Research* 49 (1989): 4640–4647.

59. Egan, T. Trees that yield a drug for cancer are wasted. *New York Times,* January 29, 1992.

60. Houston, L. L. Introduction. In *Immunotoxins,* ed. A. E. Frankel. Kluwer, 1988.

61. Vallera, D. A., and Myers, D. E. Immunotoxins containing ricin. In *Immunotoxins,* ed. A. E. Frankel. Kluwer, 1988.

62. Ehrlich, P. The relationship existing between chemical constitution, distribution and pharmacological action. In *The Collected Papers of Paul Ehrlich,* volume 1, ed. F. Himmelwaite, M. Marquandt, and H. Dale Pergamon, 1956.

63. Lambert, J. M., et al. An immunotoxin prepared with blocked ricin: A natural plant toxin adapted for therapeutic use. *Cancer Research* 51 (1991): 6236–6242.

64. King, S. R. Presentation at Rainforest Alliance Symposium, Rockefeller University, January 24, 1992.

65. Gardell, S. J., et al. Effective thrombolysis without a marked plasminemia after bolus intravenous administration of vampire bat salivary plasminogen activator in rabbits. *Circulation* 84 (1991), no. 1: 244–253.

66. Bank, N. U. Leeches, snakes, ticks and vampire bats in today's cardiovascular drug development. *Circulation* 84 (1991), no. 1: 436–437.

67. Rinehart, K. L., et al. Bioactive compounds from aquatic and terrestrial sources. *Journal of Natural Products* 53 (1990), no. 4: 771–792.

68. Klayman, D. L. Qinghaosu (artemisinin): An antimalarial drug from China. *Science* 228 (1985): 1049–1054.

69. Mann, J. Sponges to wipe away pain. *Nature* 358 (1992): 540.

70. Cassady, J. M., Gaird, W. M., and Chang, C. Natural Products as a source of potential cancer chemotherapeutic and chemopreventive agents. *Journal of Natural Products* 53 (1990), no. 1: 23–41.

71. Rinehart, K. L., et al. Marine natural products as sources of antiviral, antimicrobial, and antineoplastic agents. *Pure and Applied Chemistry* 53 (1981): 795–817.

72. Gottlieb, O. R., and Mors, W. B. *Journal of Agricultural Food Chemistry* 28 (1980): 196.

73. Schultes, R. E. Personal communication, January 24, 1992.

74. Shulka, J., Nobre, C., and Sellers, P. Amazon deforestation and climate change. *Science* 247 (1990): 1325.

75. Brody, J. E. Far from fearsome, bats lose ground to ignorance and greed. *New York Times,* October 29, 1991.

76. Terborgh, J. *Where Have All the Birds Gone?* Princeton University Press, 1989.

77. Spielman, A., and Kimsey, R. B. Zoonosis. In *Encyclopedia of Human Biology*, volume 7, ed. R. Dulbecco. Academic Press, 1991.

78. Locksley, R. M. Leishmaniasis. In *Principles of Internal Medicine*, twelfth edition, ed. J. D. Wilson. McGraw-Hill, 1991.

79. Kirchhoff, L. V. Trypanosomiasis. In *Principles of Internal Medicine* (op. cit.).

80. Berenbaum, M. Invaded by insects. *New York Times*, June 11, 1992.

81. Mitchell, C. J., et al. Isolation of eastern equine encephalitis virus from *Aedes albopictus* in Florida. *Science* 257 (1992): 526–527.

82. Epstein, P. Personal communication, October 12, 1992.

83. Telford, S. R., Pollack, R. J., and Spielman, A. Emerging vector-borne infections. *Infectious Diseases of North America* 5 (1991), no. 1: 7–17.

84. Sanford, J. P. Arbovirus infections. In *Principles of Internal Medicine* (op. cit.).

85. Steere, A. C. Lyme borreliosis. In *Principles of Internal medicine* (op. cit.)

86. Morse, S. S. Stirring up trouble: Environmental disruption can divert animal viruses into people. *The Sciences* 30 (1990), September–October: 16–21.

87. Kingman, S. Malaria runs riot on Brazil's wild frontier. *New Scientist* 123 (1989): 24–25.

88. Sanford, J. P. Arenavirus infections. In *Principles of Internal Medicine* (op. cit.).

89. Morse, S. S. The origins of "new" viral diseases. *Environmental Carcinogens and Ecotoxicity Reviews* C9 (1991), no. 2: 207–228.

90. Allan, J. S. Viral evolution and AIDS. *Journal of the National Institute of Health Research* 4 (1992): 51–54.

91. Huet, T., et al. Genetic organization of a chimpanzee lentivirus related to HIV-1. *Nature* 345 (1990): 356.

92. Hirsh, V. M., et al. An African primate lentivirus (SIVsm) closely related to HIV-s. *Nature* 339 (1989): 389.

93. Gao, F., et al. Human infection by genetically diverse SIVsm-related HIV-2 in West Africa. *Nature* 358 (1992): 495.

94. Schneider, K. Ozone depletion harming sea life. *New York Times*, November 16, 1991.

95. Roberts, L. Does the ozone hole threaten Antarctic life? *Science* 244 (1989): 288–289.

96. Bridigare, R. R. Potential effects of UVB on marine organisms of the southern ocean: Distribution of phytoplankton and krill during austral spring. *Photochemistry and Photobiology* 50 (1989): 469–478.

97. Blakeslee, S. Scientists confront an alarming mystery: The vanishing frog. *New York Times*, February 20, 1990.

98. Yoffe, E. Silence of the frogs. *New York Times*, December 13, 1992.

99. Domingo, M., et al. Morbillivirus in dolphins. *Nature* 348 (1990): 21.

100. Dold, C. Toxic agents found to be killing off whales. *New York Times,* June 16, 1992.

101. Carlsen, E., et al. Evidence for decreasing quality of semen during the past 50 years. *British Medical Journal* 305 (1992): 609–613.

102. Matlai, P., and Beral V. Trends in congenital malformations of external genitalia. *Lancet* 1 (1985): 108.

103. Angier, N. Study finds mysterious rise in childhood cancer rate. *New York Times,* June 26, 1991.

104. Kurihara, M., Aoki, K., and Tominaga, S., eds. *Cancer Mortality Statistics in the World.* University of Nagoya Press, 1984.

105. Davis, D. L., et al. International trends in cancer mortality in France, West Germany, Italy, Japan, England and Wales, and the USA. *Lancet* 336 (1990): 474–481.

106. Davis, D. L., and Hoel, D. G, eds. Trends in Cancer Mortality in Industrialized Countries. *Annals of the New York Academy of Sciences* 609 (1990).

107. Hoel, D. G., et al. Trends in cancer mortality in 15 industrialized countries, 1969–1986. *Journal of the National Cancer Institute* 84 (1992), no. 5: 313–320.

108. Stierle, A., Strobel, G., and Stierle, D. Taxol and taxane production by *Taxomyces andreanae,* an endophytic fungus of Pacific yew. *Science* 260 (1993): 214–216.

109. Altma, L. K. Sharks yield possible weapon against infection. *New York Times,* February 15, 1993.

110. Chren, M. M., and Bickers, D. R. Dermatological pharmacology. In *The Pharmacologic Basis of Therapeutics* (op. cit.).

Glossary

acinar
a term used interchangeably with *alveolar* to refer to the smallest sacs in an organ, such as the lung, that combine in grape-like clusters to form lobules.

acute lymphocytic leukemia
leukemia associated with overgrowth of the lymphoid tissue, in which lymphocytes and their precursors, lymphoblasts, predominate.

acute myelocytic leukemia
leukemia arising from myeloid (blood-forming) tissue in which polymorphonuclear leukocytes (white blood cells) and their precursors predominate.

acyclovir
a medicine with selective anti-viral activity against *herpes simplex* virus.

adipose tissue
connective tissue composed of fat cells.

African sleeping sickness
a disease, caused by a protozoan and carried by tsetse flies in tropical Africa, that infects the central nervous system and which in untreated cases is generally fatal.

alveolar cells
cells that line alveoli in the lung, the tiny balloon-like sacs at the end of bronchioles where the transfer of oxygen and carbon dioxide between air and blood occurs.

Alzheimer's disease
a common primary degenerative disease of the brain characterized by a loss of intellectual functions.

amyotrophic lateral sclerosis (Lou Gehrig's disease)

a progressive degenerative disorder of motor neurons, manifested clinically by muscular weakness atrophy and eventual complete paralysis.

angiosarcoma of the liver

a rare malignant tumor of the liver arising from endothelial and fibroblastic tissue (types of connective tissue) and associated with exposure to vinyl chloride monomer, arsenic, and Thorotrast (a contrast agent used in x-ray studies).

angiotensin

a protein (composed of multiple amino acids) in the blood that is a powerful elevator of blood pressure and stimulator of the secretion of aldosterone (a steroid hormone from the adrenal cortex that regulates sodium, potassium, and chloride metabolism).

antimetabolite

a substance that can replace an essential metabolite in a chemical reaction, such as carbon monoxide replacing oxygen in combination with hemoglobin.

anti-oxidant

a substance that delays or prevents oxidation.

arbovirus

a virus transmitted by an arthropod.

aromatic amino acids

amino acids containing a benzene or a quinonoid ring.

arthropod

an organism belonging to the phylum *Arthropoda,* which includes insects, arachnids, and crustaceans.

atrial fibrillation

a cardiac arrhythmia characterized by rapid, randomized contractions of the atria (the small chambers of the heart), causing an irregular heart rate.

atrial flutter

a cardiac arrhythmia in which the atrial contractions are rapid (200–320/minute) but regular.

autoimmune

an immune response to antigens from one's own tissues.

AV nodal tachycardia

a cardiac arrhythmia caused by an irritable focus at the atrio-ventricular node resulting in a rapid heartbeat.

bacterial endocarditis

a bacterial inflammation of the endocardium, or lining of the heart chambers, often involving the valves.

basal metabolic rate

the energy required to maintain the body's metabolic processes in a resting state, 14–18 hours after eating.

becquerel

a unit of radioactivity equal to one nuclear disintegration per second. A disintegration is the transformation of a nucleus, either spontaneously or by interaction with radiation, in which particles or photons are emitted.

biota

the sum total of all life on Earth.

bradykinin

a small protein that is an extremely potent vasodilator (resulting in a drop in blood pressure) produced by the activation of the kinin system in a variety of inflammatory conditions.

bronchiolitis

a viral infectious disease of the bronchioles (the smallest airway tubes in the lung), seen in young children.

bronchospasm

spasmodic contraction of the walls of the bronchi, the airways leading from the trachea to the lungs, making it difficult to breathe.

buffering

the neutralizing of acids or alkalis without affecting the pH value.

byssinosis

a chronic bronchitis and emphysema with fibrosis resulting from exposure to cotton or linen-fiber dust.

catabolism

the process of metabolism involving the breaking down of complex molecules into simpler ones with a release of energy.

carcinogenic

causing cancer.

cardiac arrhythmia

any variation from the normal rhythm of the heartbeat.

Chagas' disease

a vector-borne disease of Central and South America caused by the protozoan *Trypanosoma cruzi* and resulting in damage to the heart and the gastrointestinal tract.

cholinesterase

an enzyme, found in the blood serum, the liver, and the pancreas, that catalyzes the breakdown of choline esters, such as acetylcholine. Organophosphate insecticides poison this enzyme.

clot lysis time

the time required for the breakdown of a blood clot.

cohort mortality studies

studies of the mortality of a group of individuals identified by a common characteristic, such as age.

communicable disease

a disease capable of being transmitted directly or indirectly from person to person, through an intermediate host or vector, or via the environment.

curie

a measure of radioactivity. One curie is the rate of disintegration of one gram of radium: 37 billion disintegrations per second.

cytotoxic

toxic to cells.

deciliter

one-tenth of a liter.

dengue fever

an acute viral infectious systemic disease characterized by extreme prostration and severe muscle, bone, and joint pains and found in tropical and subtropical regions. Spread by various species of mosquitoes, dengue can result in hemorrhagic fever which carries a high mortality in untreated cases.

dermal

pertaining to the skin.

dioxin

a carcinogenic, teratogenic, and mutagenic chlorinated hydrocarbon formed at incineration sites at high temperatures from chlorine and hydrocarbon raw materials. Dioxin is also a contaminant of 2,4,5-T, a widely used herbicide known as Agent Orange.

dissociation curve of oxyhemoglobin

the curve that defines the relationship between oxygen concentration and amount of oxyhemoglobin in red blood cells at a specified temperature and pressure.

dracunculiasis

an infection caused by *Dracunculus medinensis,* a nematode also known as the guinea worm.

dyspnea

difficult or labored breathing.

eclampsia

a condition occurring in the mother between the twentieth week of pregnancy and one week after delivery and characterized by seizures not due to epilepsy, cerebral hemorrhage, or other cerebral conditions.

encephalitis
inflammatory disease affecting the brain.

encephalopathy
refers to any defect of the structure or function of brain tissues, including those caused by systemic diseases such as hypertension or metabolic disorders or by toxic agents. Toxic encephalopathy may develop acutely after high-dose exposure or over time after low-dose exposure.

endemic
present at all times in a particular population or geographic area.

filariasis
a group of diseases caused by infection with nematodes and transmitted by mosquitoes. The most common type is due to *Wuchereria bancrofti,* which invades the lymphatic vessels and in late stages may lead to elephantiasis.

gigatonne (Gt)
one billion tonnes.

glycoprotein
a protein compound with a carbohydrate group.

half-life
the time required for a radioactive substance to lose 50% of its radioactivity by decay.

halogenated cyclic compound
a cyclic organic compound containing fluorine, chlorine, bromine, or iodine.

hectare
a unit of area equal to 10,000 square meters, or 2.471 acres.

Hodgkin's disease
a malignant lymphoma characterized by a painless progressive enlargement of the lymph nodes, the spleen, and other lymphoid tissue.

host
an organism that harbors and provides nourishment for an infectious organism.

hydraulic gradient
pressure differences that govern the way fluids move through a medium.

hyperkeratosis
overgrowth of the stratus corneum (the other layer of the skin, consisting of dead, scaly cells).

hypoxemia
a reduction in oxygen carried by the blood, leading to levels inadequate for normal tissue respiration.

incubation time

the interval of time between the entrance of an organism into a host and the onset of symptoms caused by infection.

indomethacin

a nonsteroidal anti-inflammatory agent widely used in the treatment of arthritis.

in vitro

occurring "in glass" in a laboratory or some other artificial environment.

in vivo

occurring "in life" in an intact organism.

isomers

compounds having the same chemical composition and weight but a different arrangement of atoms.

isotopes

nuclides (species of atoms) having the same number of protons in their nuclei (and therefore the same atomic number) but differing in the number of neutrons (and therefore in atomic weight).

Kaposi's sarcoma

a malignant cancer involving blood-vessel cells, usually producing bluish-red nodules in the skin and seen especially in patients with compromised immune systems, such as those with AIDS.

keratinocyte

the skin cell (comprising 95% of the total cells in the epidermis) that synthesizes keratin, a very insoluble protein that forms the hair, the nails, and the outermost horny layer of the skin.

Langerhans cells

cells in the pancreas that synthesize, store, and release the hormones insulin, glucagon, and gastrin.

leukemia

a progressive malignant disease of the blood-forming organs, characterized by uncontrolled proliferation of immature and abnormal leukocytes (white blood cells) and their precursors in the blood and the bone marrow.

linear dose response

a straight-line relationship between the dose of a substance and a particular response.

lymphocyte

a mature white blood cell, found in lymphatic tissue, that mediates the body's cellular and humoral (antibody-producing) immune response.

macrophages

large phagocytic cells found in connective tissues, especially where there is inflammation.

mesothelioma

a malignant cancer arising from mesenchymal cells found in the lining of the lung (the pleura) and the gastrointestinal tract (the peritoneum). Approximately 85% of patients with mesothelioma have a history of occupational or environmental asbestos exposure.

methemoglobinemia

the presence of more than 1% methemoglobin in the blood, resulting in cyanosis. Methemoglobin, formed when hemoglobin is oxidized from the ferrous to the ferric state, does not release carried oxygen to the tissues.

μg

microgram (one millionth of a gram).

mutagenicity

the capability of inducing genetic mutation.

myasthenia gravis

an autoimmune disorder of neuromuscular function characterized by exhaustion of muscle groups or (more rarely) of the entire skeletal musculature.

mycotoxin

a toxin derived from fungi.

myoglobin

muscle hemoglobin, which functions to carry and store oxygen in muscle.

nanometer

one billionth of a meter.

nematicides

agents that are toxic to nematodes.

neuroblastoma

a tumor, often with widespread metastases composed of primitive neural tissue, affecting mostly infants and children.

nitrogen fixation

the conversion of atmospheric nitrogen, by bacteria that live in the soil, into nitrogen compounds that can be used by green plants.

nitrosamines

any of a group of compounds that are formed by the combination of nitrates and amines.

nitrosating agents

chemical agents that produce nitroso compounds.

nitroso compounds
chemicals containing the group —N=0.

non-Hodgkin's lymphoma
a malignant lymphoma that is similar to Hodgkin's disease but more widespread.

nuclide
a species of an atom capable of existing for a measurable period of time and defined by the numbers of protons and neutrons and by the energy content of the nucleus.

opportunistic infection
an infection caused by organisms that ordinarily do not cause disease but which do so under certain conditions, such as with impaired immune responses.

osteoporosis
a chronic disease in which there is reduced bone mass and density because of increased bone resorption, resulting in fractures.

peripheral neuropathy
a disease process characterized by degeneration of the peripheral nerves, usually due to some toxic, metabolic, or vascular disturbance.

phagocytosis
the ingestion of microorganisms, cells, and foreign particles by cells of the reticuloendothelial system known as *phagocytes*.

pharmacokinetics
the way drugs in the body are absorbed, distributed, metabolized, and excreted over time.

pheromones
hormonal substances released by an organism that affect the development, the reproduction, and/or the behavior of other members of the same species, usually of the opposite sex.

pneumoconiosis
a group of diseases caused by the permanent deposition of particulate matter in the lungs and by the tissue's reaction to this matter.

polyarthritis
inflammation affecting a number of joints.

ppm
parts per million by weight—for example, 1 milligram of a substance contained in 1,000 grams (1 kilogram) of a solution.

pyrethrin
a naturally occurring plant insecticide.

pyrimidine

an organic nitrogenous base consisting of a heterocyclic compound. The pyrimidines cytosine and thymine pair with the pyrines guanine and adenine in all DNA.

rad

"radiation absorbed dose." One rad is the amount of ionizing radiation that deposits 100 ergs of energy in each gram of exposed tissue. One gray is equal to 100 rads.

radiation

the production and transmission of energy in the form of electromagnetic waves or particles.

radioactivity

the property of the nuclei of certain atoms to emit radiation spontaneously in the form of gamma rays, alpha particles, or beta particles.

radionuclide

a radioactive nuclide (species of an atom).

Raynaud's phenomenon

attacks of pallor and cyanosis of the extremities on exposure to moderate cold.

rem

"roentgen equivalent man," a unit of ionizing radiation equivalent to the dosage that produces the same amount of tissue-damage as one roentgen of high-voltage xrays. *Rem* is often used interchangeably with *rad,* but they are not identical. A high rad dose of external alpha radiation (which is not penetrating) would cause little tissue damage, and would therefore be lower on the rem scale. The same dose internally would cause great damage and have a high rem value. The maximum permissible annual dose for an individual is 500 millirems; for those who work in radiation occupations, it is 5 rems. The exposure unit for considering health effects in a population is the person-rem, the product of the number of people and the total rem exposure. A sievert is equal to 100 rems.

renin

an enzyme synthesized in the kidney and released in response to various metabolic stimuli. Renin converts angiotensin to angiotensin 1.

rheumatoid arthritis

a chronic systemic disease characterized by inflammation of the joints.

schistosomiasis (bilharziasis)

an infection caused by flatworms of the *Schistosoma* genus in tropical and subtropical regions, having freshwater snails as intermediate hosts, and causing damage to the liver, the spleen, and sometimes the bowel, the bladder, and the kidneys.

silicosis

a pneumoconiosis caused by inhalation of particles of silica, quartz, or slate.

suppressor T cells
differentiated T lymphocytes that inhibit the response of other lymphocytes to an antigen.

tachypnea
excessively rapid breathing.

teratogenicity
the tendency of a substance to produce abnormalities of formation (or physical defects) of a developing embryo.

thrombotic
forming a blood clot.

vasodilating
increasing the caliber of the small arteries.

vasospastic
characterized by spasmodic constriction of the small arteries.

vector
an insect or other organism that carries an infective agent from one host to another.

yellow fever
an acute infectious disease of tropical and subtropical regions, caused by a virus and transmitted to man by mosquitoes. In severe cases there is necrosis of the liver and jaundice.

Volume Editors, Contributors, and Consultants

Volume Editors

Eric Chivian, M.D.

Assistant Clinical Professor of Psychiatry, Harvard Medical School. Psychiatrist, Massachusetts Institute of Technology. Revived Physicians for Social Responsibility in 1978 (with Drs. Ira Helfand and Helen Caldicott). Co-founder of International Physicians for the Prevention of Nuclear War, recipient of the 1985 Nobel Peace Prize.

Michael McCally, M.D., Ph.D.

Lecturer in Medicine, Pritzker School of Medicine, University of Chicago. Health Program Officer, Chicago Community Trust. Former Professor of Public Health and Preventive Medicine, Oregon Health Sciences University. Former Chief, Environmental Medicine Division, U.S. Air Force Aerospace Medical Research Laboratory.

Howard Hu, M.D., M.P.H., Sc.D.

Assistant Professor of Environmental Health, Harvard School of Public Health. Instructor in Medicine, Harvard Medical School. Director, International Commission to Investigate the Health and Environmental Consequences of Nuclear Weapons Production, International Physicians for the Prevention of Nuclear War.

Andrew Haines, M.D.

Professor and Head of Department of Primary Health Care, University College London Medical School. Director of Research and Development, North East Thames Regional Health Authority. Vice-President, MEDACT (Medical Action for Global Security).

Contributors

Dean B. Baker, M.D., M.P.H.

Associate Professor of Community Medicine, Mount Sinai School of Medicine.

Elizabeth L. Bowen, M.D., Ed.D.

Assistant Professor of Community and Preventive Medicine, Morehouse School of Medicine. President, Physicians for Social Responsibility.

David C. Christiani, M.D., M.P.H.

Associate Professor of Environmental Health, Harvard School of Public Health. Associate Professor of Medicine, Harvard Medical School.

Anthony Cortese, Sc.D.

Environmental consultant. Former Dean of Environmental Programs, Tufts University. Former Commissioner, Massachusetts Department of Environmental Protection.

Ira Helfand, M.D.

Emergency Room Physician, Cooley Dickinson Hospital, Northampton, Massachusetts. Revived Physicians for Social Responsibility in 1978 (with Drs. Eric Chivian and Helen Caldicott). Treasurer of PSR.

Nancy K. Kim, Ph.D.

Director, Division of Environmental Health Assessment, New York State Department of Health. Associate Professor of Environmental Health and Toxicology, State University of New York at Albany.

Philip J. Landrigan, M.D.

Professor and Chairman, Department of Community Medicine, Mount Sinai School of Medicine. Former Director, Division of Surveillance, Hazard Evaluations, and Field Studies, National Institute of Occupational Safety and Health.

Alexander Leaf, M.D.

Distinguished Physician, VA Affairs Medical Center, Brockton and West Roxbury, Massachusetts. Jackson Professor of Clinical Medicine (Emeritus), Professor and Chairman (Emeritus), Departments of Medicine and Preventive Medicine, Harvard Medical School.

Jennifer Leaning, M.D.

Instructor in Medicine, Harvard Medical School. Medical Director, Harvard Community Health Plan. Editor-in-Chief, *PSR Quarterly.*

Kenneth A. Lichtenstein, M.D.

Associate Clinical Professor of Medicine, University of Colorado Health Sciences Center. Past President, Physicians for Social Responsibility.

Consultants

Lincoln C. Chen, M.D., M.P.H.

Takemi Professor of International Health, Harvard School of Public Health. Director, Center for Population and Development Studies, Harvard University.

Anne Ehrlich

Research Associate, Department of Biological Science, Stanford University

Paul Epstein, M.D., M.P.H.

Clinical Instructor In Medicine, Harvard Medical School

Rodney M. Fujita, Ph.D.

Senior Scientist, Environmental Defense Fund

Stephen S. Morse, Ph.D.

Assistant Professor of Virology, Rockefeller University.

S. Jay Olshansky, Ph.D.

Research Associate, Department of Medicine and Population Research Center, University of Chicago.

Richard Evans Schultes, Ph.D.

Professor of Biology and Director of Botanical Museum (Emeritus), Harvard University.

Andrew Spielman, Ph.D.

Professor of Tropical Public Health, Harvard School of Public Health

Abdullah Toukan, Ph.D.

Science Advisor to His Majesty, King Hussein of Jordan.

Arthur Westing, Ph.D.

Adjunct Professor of Ecology, Hampshire College.

Index

---

food contamination from, 50–56
water pollution from, 39–40
Pest management, integrated, 53
Petroleum products, in water, 41–42
Pharmacokinetic models, 9
Photochemical oxidants, 19–20
Photokeratis, 146
Physicians
and population growth, 173, 187–188
role of, 9–10
skills needed in, 5–6
training of, 5
Phytoplankton, 142, 148
Pipes, lead in water from, 36–38
Pit vipers, 202–203
Plutonium, 116
Polyarthritis, 159
Polybrominated biphenyls (PBBs), 57
Polychlorinated biphenyls (PCBs), 57
Polycyclic aromatic hydrocarbons (PAHs), 23
Polyhalogenated aromatic hydrocarbons, 57
Pope, Alexander, 193
Population growth, 171–173
aging and, 182
in cities, 181
climate and, 167
demographic transition in, 176
demographic trap in, 176
and desertification, 179
and food supplies, 178–179
maternal mortality and, 180–181
momentum of, 177–178
overcrowding from, 181–182
physicians and, 173, 187–188
policies and, 186–187
rate of, 174–175
refugees from, 182
and species extinction, 195
stabilization of, 182–186
vs. technology, 179
teen pregnancy and, 180
war and, 130–131
Prostaglandins, 202
Proteinuria, 61
Pseudopterosins, 202

Psoralens, 207–208

Quinine and quinidine, 204–205

Radiation, 93–94. *See also* Electromagnetic radiation; Ultraviolet radiation
assessing health effects of, 110–111
biology of, 94–96
epidemiologists and, 97–99
genetic damage from, 99
linear, no-threshold hypothesis of, 96
new information on, 99–100
from nuclear power, 111–116
supralinear hypothesis of, 96
from weapons, 100–110, 132
Radioactive fallout, and food, 56
Radioactive substances, and water, 42
Radioactive wastes, transuranic, 115
Radionuclides, in food, 56
Radon, in water, 42–43
Rain
acid, 17, 62, 217
and climatic changes, 165
Rain forests, 3
infectious diseases and, 213
species extinction and, 195
war and, 130
Ramazzini, B., 13
Refugees
from population growth, 182
from war, 125
Renewable resources, and population, 177
Respiratory problems. *See also* Air pollution
climate and, 161–162
ultraviolet radiation and, 148
Ricin, 208–209
Rift Valley fever, 160
Risks
analysis and assessment of, 6–9
characterization of, 7
communication of, 9
models for, 8
from radiation, 102, 110–111
Ross River virus, 159
Rosy periwinkle, 205
RU 486 pill, 186